ECOLOGY OF HIGHLANDS

MONOGRAPHIAE
BIOLOGICAE

Editor

J. ILLIES

Schlitz

VOLUME 40

Dr W Junk bv Publishers The Hague–Boston–London 1980

ECOLOGY OF HIGHLANDS

M. S. MANI
&
L. E. GIDDINGS

Dr W Junk bv Publishers The Hague–Boston–London 1980

Distributors:

for the United States and Canada

Kluwer Boston, Inc.
190 Old Derby Street
Hingham, MA 02043
USA

for all other countries

Kluwer Academic Publishers Group
Distribution Center
P.O. Box 322
3300 AH Dordrecht
The Netherlands

Library of Congress Cataloging in Publication Data   **CIP**

Mani, M S
  Ecology of highlands

  (Monographiae biologicae; v. 40)
  Bibliography
  Includes index.
  1. Mountain ecology. I. Giddings, Lorrain E., joint author. II. Title. III. Series.

QP1.P37 vol. 40 [QH541.5.M65] 574s [574.5′264] 80-14798

ISBN-13: 978-94-009-9176-7   e-ISBN-13: 978-94-009-9174-3
DOI: 10.1007/978-94-009-9174-3

Table of Contents

List of figures and plates

Preface

High altitude research is comparatively a recent development. With the notable exceptions of entomology, botany and perhaps some aspects of human acclimatization, our knowledge of high altitude environment is extremely scanty. There is at present no comprehensive handbook on the general ecology of highlands. This book aims at providing such a text for the use of the general scientist, engineer, biologist and university students. It summarizes and critically reviews current developments and focusses attention on urgent problems of highland ecology needing future studies.

This book has grown out of our explorations and experiences in the highlands of Asia and South America. The results of explorations of the high altitude plants and insects on the Himalaya, Alai-Pamirs, Central Tien Shan and Caucasus by one of us (MSM), discussed in earlier publications, are reviewed here in the light of recent advances. Many years' experience of teaching and research in the University of Sucre (3400 m) and studies at Potosi (4000 m), La Paz and Chacaltaya (5000 m) in Bolivia by the second author (LEG) cover problems in physical chemistry, meteorology, engineering and other physical aspects of highland environments.

The book contains fifteen chapters. The first chapter explains the modern concept of high altitude and draws special attention to the outstanding peculiarities of highland ecosystems, which serve to distinguish them from other ecosystems. Particular emphasis is laid on the fact that highlands are basically regions of semi-attenuated atmosphere. Some of the major highlands regions of the world are briefly mentioned. In line with the practical orientation of this book, the second chapter presents the atmospheric pressure and temperature of highlands and shows how to estimate their values. It shows how other properties of the air depend on these two factors and also discusses the problems of their estimation. The third chapter treats of the radiation environment in highland regions as a direct result of rarified atmosphere. There is a brief discussion of the methods of estimating the solar radiation in relation to atmospheric haze and altitude. In the next four chapters the practical problems that face workers in highland laboratory are discussed. Correcting boiling points for the reduced atmosphere at high altitude may appear to be simple, but most charts available at present are not suitable for use at elevations above 3000 m. Chemical effects of the reduced pressure are generally indirect and for this reason treacherous. The practical problems of manometry and psychrometry are then discussed in some detail. The problems which face the engineer in highlands are often quite complex. The eighth chapter is devoted to a brief outline of the more vexing ones for the engineer, with particular emphasis on combustion and azeotropy. Remote sensing of the highland environments from aeroplanes and artificial satellites may be the most powerful tools available to the student of highland ecology. The ninth chapter presents a very practical orientation to this field, suggesting what can be learnt about highlands from space.

The outstanding characters of the highland vegetation are briefly reviewed in Chapter X, with suitable illustrations mainly from the Pamirs-Himalaya and the Andes. The next chapter reviews our knowledge of animal life, with particular emphasis on their ecological specializations. Three chapters are devoted to man in highland ecosystems: the first outlines the peculiarities of the natives of highlands and is the result of collaboration with Dr A. Pardo, Rector, Sucre University, Bolivia. This is followed by a review of the physiological-pathological changes induced in the sea-level resident exposed to high altitude conditions. A whole chapter is devoted to the problems of human acclimatization to high altitudes. The last chapter of the book attempts to focuss attention on some of the problems of highland ecology, which require research in future and also draws attention to the impact of man on highland ecosystems.

In writing this book we are painfully aware of many questions for which we have now no answer. Our experience has shown that it is much harder to be a scientist or engineer at high altitude than in most lowland areas. One must not only overcome the ordinary problems of scientific or engineering disciplines, as elsewhere, but must also overcome isolation from the main-stream of science and lack of equipment built for use under highland conditions. He must also compensate for the unique problems caused by altitude, most of which have never been satisfactorily discussed before. If this book makes life easier for highland scientists and stimulates further research on various aspects of highland environments, we would have achieved our purpose.

We take this opportunity of expressing our sincere thanks to our students and collaborators in the School of Entomology, St. John's College, Agra and in the University of Sucre, Bolivia, for valuable help in the course of our work. We are specially indebted to Dr Antonio Pardo Subieta, Director, Faculty of Research, University of Sucre, Bolivia, for constant encourage-ment and for active collaboration in writing the twelfth Chapter. We are also grateful to Dr Rosendo Carreras of the Faculty of Chemical Engineering, University of Sucre and to Dr P. Rafael Queralt Teixido, Editor of the *Afinidad*, Spanish language journal of chemistry, for useful advice and numerous courtesies.

M. S. MANI
L. E. GIDDINGS

Addresses of authors

M. S. Mani, School of Entomology, St John's College, Agra 282002, India.
L. E. Giddings, Instituto Nacìonal de Investìgacíones Sobre Recursos Bióticos, Apdo Postal 63, Xalapa, Veracruz, Mexico.
Dr A. Pardo, Rector, Sucre University, La Paz, Bolivia.

I Introduction

M. S. Mani

1. What is Highland?

The highland is a high altitude region, which is basically an area of relatively low atmospheric pressure (3, 4, 5).* The reduced atmospheric pressure of high altitude is associated with atmospheric cold and aridity, deficiency of oxygen and carbon dioxide, intense insolation and rapid radiation, high ultraviolet and other effects as chain reactions. These effects become significant and the general environment, vegetation and animal life become markedly different from those of lowland areas, the boiling point of water is so low that cooking and many chemical reactions can only be carried out with difficulty, breathing is difficult and muscle fatigue great at elevations of about 3000 m above mean sea level, (1, 4, 5). By our normal standards the conditions are severely extreme. All areas above this altitudinal limit are recognized as highlands.

Highlands occur on mountains, plateaus and altiplano. The terms mountains and hills are only relative and cannot be precisely expressed in units of altitude. A hill is defined, for example, as a high mass of land, less than a mountain and a mountain is a high hill. Their use is often more or less arbitrary. Mount Washington in the eastern United States is, for example, not so high as the Black Hill, but it is called, by general consent, a mountain. Again, the Rocky Mountains of America are nowhere so high as most of the hills in India. An essential feature that distinguishes a mountain from a plateau is the relatively limited width at the summit. The altiplano is extensive, almost level, elevated 'plain' on the South American Andes.

Mountain masses do not generally stand alone, but form parts of irregular groups (e.g. the Adirondack Mountains of USA) or of large regular belts, extending more or less unbroken over vast areas, often the entire length of a continent, and constituting ranges, system and chains. A mountain range is a complex or a series of related and more or less continuous ridges. A mountain system is a group of mountain ranges of similar form, structure, alignment and origin. We have, for example, the Alpine-Himalayan Systems. The expression mountain chain is applied to any elongate mountain unit of several groups and systems, regardless of similarity and relationship. The series of mountain chains, systems and ranges that form a more or less compact elevated area of vast expanse in America is called *cordillera*.

Broadly speaking, mountains are of two types, (a) tectonic or original mountains and (b) the relict or subsequent mountains. The tectonic mountains are the result of piling up of material at the surface the earth or of subterranean action leading to folding and rupturing of the crust. The relict mountains are merely residual portions of former elevated tracts, which

* Numbers in brackets refer to references at end of each chapter.

have been gradually reduced in extent and largely subdued by forces of decay. The tectonic mountains are of two major groups: (i) the accumulation mountains and (ii) the deformation mountains. The former group includes the volcanic and epigene types. The deformation mountains are subdivided into the (i) fold mountains, (ii) the dislocation mountains and (iii) the laccolith mountains.

Volcanoes consist of material ejected from and accumulated around an orifice in the crust. Some of them are composed entirely of rocks, others of sheets of masses of lava and still others of debris and lava. Volcanoes, which often rise to great elevations, are of three types, extinct, dormant and active. The Chilean Aconcagua and the Mexican Ixtaccihuatl are peaks of volcanic origin. Mount Kilimanjaro and the Chimborazo are typical extinct volcanoes. In the Cascade Range of the USA are found some typical dormant volcanoes. Nearly four hundred active volcanoes, known at present, are situated mostly close to the sea in regions, where the folding and faulting of the crust are still active. All great mountain ranges and chains are fold mountains, which consist of much folded and steeply inclined rock strata. They owe their origin either to one long lateral thrust or to two or more repeated thrusts, separated by intervals of time, so that the later folds flank the earlier ones. Some of the dislocation mountains also constitute chains and ranges, but do not equal the great fold-mountain systems like the Alps, the Himalaya or the Andes. The dislocation mountains arise by the fracturing of the crust, by the unequal subsidence of the ground, along lines of vertical displacement. Basically they are segments of the crust, which have maintained their relative positions, while the neighbouring tracts have broken away and subsided. The parallel ranges of the Great Basin of North America are perhaps one of the best known dislocation mountains. We have here a plateau, surrounded on all sides by lofty mountains, extending north–south some 1300 km and east–west about 800 km between the Sierra Nevada of California and the Wahsatch Mountains. The laccolith mountains are the result of subterranean igneous action, the crust being bulged upwards, owing to the pressure of concealed mass of molten rock from below. The Henry mountains of Utah are typical laccolith mountains.

2. The High Altitude Environment

The outstanding characters of high altitude environments have recently been described, with special reference to the basic differences between the high altitude and lowland environments (4, 5). The high altitude environment is essentially an environment of progressively thinner atmosphere. The gravitational settling of the molecules of air near sea level results in a gradual fall in the density of air and with it also of the atmospheric pressure as the altitude increases (Chapter II). It is not only the partial pressures of the component gases like nitrogen, oxygen and carbon dioxide, but also of the water vapour and the density of aerosol that progressively fall with the increase in altitude (Chapters VI, VII). The lowered density of air at high altitude results in increase of the transparency of the air, thus modifying the conditions of insolation, absorption and radiation and secondarily evaporation. Owing to the relative thinness of the air, only a small part of the solar radiation is directly absorbed by the air. At the same time, almost all of the terrestrial radiation is allowed to escape, so that the well known 'greenhouse effect', with which we are familiar in the lowland regions, is nearly absent at high altitudes. The atmospheric temperature (in the shade) falls, therefore, with increase in altitude, in accordance with what is known at the 'lapse rate' (chapter II). The altitude at which the mean annual air temperature in the shade does not rise above 10°C is the timberline. The altitude above the timberline, at which the air temperature similarly does not rise above the freezing point of water even in summer, is the snowline. Regions at altitudes higher than the snowline are characterized by the persistence of the snow cover throughout the year; the winter snow does not melt away completely. The difference between the air temperatures in the shade and in direct sunshine is greatly accentuated by the increased air transparency at high altitude. The difference between the shade and sun temperatures thus tends to increase with increase in altitude. This difference may range, for example, from 35°C to as much as 73°C (4, 5). Regardless of the prevailing low atmospheric temperatures in the shade, most objects exposed to the direct sun at high altitudes on mountains, thus become warmed up far more rapidly than would be the case at sea level. The thin air and the scanty water vapour present in the high altitude atmosphere favour the rapid desiccation of bodies exposed to the air and to bright sunshine. It is generally found that the mean evaporating power of air at high altitudes is mostly higher than should be expected at the prevailing air temperature.

All environmental factors are mutually interactive; independent action of any single factor is unknown in Nature. The effect of any individual factor is very profoundly modified by the action of a number of other closely interlocked and mutually adjusting factors. Not only does each environmental factor act separately on the high altitude organisms, but the sum total of all the factors complex is by itself, in its turn a factor. In the effective factor complex of the high altitude environment, each factor exerts its own influence but always in a definite interaction with every other factor. In all cases there is neither a simple alternation nor even a succession of the individual factors, but each factor works directly or indirectly on all others. Through this interaction it affects the organisms in a most complex manner. All the environmental factors are thus closely and inseparably interlinked and intricately intertwined in a complex chain of cause and effect. In Nature there is, therefore, but a single, indivisible and complex factor, viz. the *high altitude environment*, an indivisible and integrated unit, in which we cannot isolate any factor element from the whole complex. The entire complex of the closely integrated and interlinked factors constitutes in reality a self-regulating dynamic system with organisms, which we recognize as the *high altitude ecosystem*.

All highlands are treeless areas, above the altitudinal limits of closed forest (timberline). The elevated areas above timberline are generally called alpine-arctic zones. Actually the alpine and arctic zones are only very small parts of high altitude ecosystem (Fig. 1-1). This zonation is also based on erroneous application of the well known bioclimatic law. According to this law (2), a high altitude region of about 1500 m above mean sea level would correspond to the conditions of atmospheric temperature at sea level in the latitude 12° 30' further to the north. There is also a general lag of seasonal events for each rise of elevation of about 122 m, corresponding to about one degree latitude in the north. Most workers seem to be so greatly impressed by this apparent parallelism that they have fallen into the serious error of equating the general climatic conditions at high altitudes with those prevailing at high north latitudes in the tundra and arctic regions. These well known zones are no doubt more or less valid in a general way as one proceeds from the tropics to the high northern latitudes in the plains, but are not wholly satisfactory on highlands. There is in this case the risk of overlooking a most important ecological factor that sharply distinguishes the *high altitude* from the *high latitude* environments. The rôle of only the atmospheric temperature, isolated from other ecological factors, is over emphasized. *The high altitude is not only a progressively cold region but is primarily a region of gradually semi-attenuated atmosphere: this is by no means the case with high latitude (arctic) regions.* The temperate-arctic zones of north latitudes are basically *lowland* areas of moderate or low temperatures *at dense atmosphere,* but the high altitude is a region of low temperature of *gradually thinner atmosphere.* The cold environments of high altitudes and of the north and south latitudes arise, however, from fundamentally different effects. The atmospheric cold of the subarctic, arctic and antarctic environments is conditioned not by the inadequate warming of a relatively thin or semi-attenuated atmosphere by the solar radiation, as is the case at high

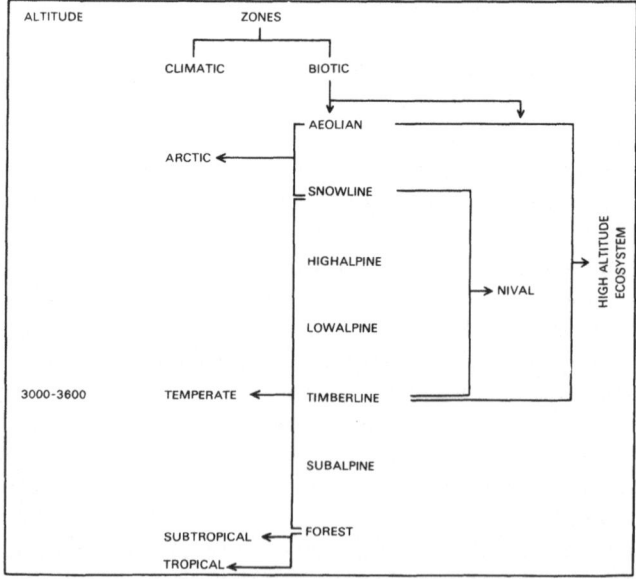

Fig. 1-1. Climatic and biotic zones on mountains comprising the high altitude ecosystem.

altitudes, but entirely by the difference in the angle of incidence of this radiation. The atmospheric cold of high latitudes is the result of relatively fewer sunrays incident on unit surface. At high altitude, on the other hand, the atmospheric cold is the result of actually less air being warmed up, although there is in reality more sunrays per unit area of surface than at high latitudes. While certainly the mean atmospheric pressure and with it the density of the air are known to be influenced to some extent by the differences in latitude (Chapter VI), even deep within the arctic or the antarctic regions the atmosphere is by no means as thin as even on a moderately high mountain. *High altitude cold is conditioned by precisely the same factors and is also more related to the cold of the aerospace environment and of the interplanetary space rather than to the arctic cold. In a sense the high altitude is a transition to the aerospace.*

The high altitude environments of highlands in different parts of the world present characteristic differences. The environmental conditions, which may be attributed to altitude, are more or less profoundly modified by the massiveness of the mountain system, the trendline of the mountain ranges and the general climate of the surrounding lowland country. The high altitude environment, even at comparable elevations, is thus often different on different mountains, situated in different climatic regions of the world.

As is well known, tropical regions are generally characterized by the absence of sharply defined temperature-conditioned seasons. The north temperate and the north cold regions are large continental landmasses, with typically continental climate, characterized by sharp temperature-conditioned seasons. In the southern temperate and cold regions, we find relatively narrow meridional land areas or numerous scattered islands, surrounded by extensive oceans, so that the general climate is more oceanic and nearly isothermal conditions prevail the whole year round. These differences in the general climatic conditions of the lowlands very profoundly modify the high altitude environments of mountains situated in different regions. The environment at high altitudes on equatorial and tropical mountains is remarkable for the dominance of diurnal temperature fluctuations over the seasonal fluctuations, so that there is no sharp difference between summer and winter. We find instead nocturnal sub-zero temperatures and frost every day, almost throughout the year at sufficiently high altitude. Nocturnal frost on tropical mountains may indeed occur even at elevations of 1500 m, depending on the general weather conditions, topography, vegetation, etc. Cloudy days prevent insolation and if they alternate with windless and starlit nights, the cold air caused by re-radiation is unlikely to stagnate and be cooled further, especially if the vegetation cover is low as turf, to give rise to frost. The high altitude environment on the mountains of the north temperate and boreal cold regions is characteristically seasonal-thermic, which marked differences in the mean winter and summer atmospheric temperatures. The high altitude environment of the mountains of the southern hemisphere is typically isothermic.

The characteristic differences of the high altitude environments in different climatic regions of the world are summarized in Table 1-1.

Table 1-1. Highland environments in different climatic regions of the world.

Character of environment	Mountains in north latitudes	Equatorial or tropical Mountains	Mountains in southern latitudes
General climate	Continental; seasonal-thermic; seasonal temperature fluctuations are greater than the diurnal fluctuations	Diurnal-thermic; tropical; diurnal temperature fluctuations are greater than the season fluctuation	Oceanic; isothermic neither the seasonal nor the diurnal temperature fluctuations are pronounced
Atmospheric cold	Due both to the thin air and the slanting incidence of sunrays	Due only to the thin air	Due both to the thin air and the slanting incidence of sunrays
Insolation	Intense due to the thin air	Intense due to both the thin air and vertical sunrays	Due only to the thin air
Winter snow cover	Pronounced	No superabundant winter snow; nocturnal frost daily almost the whole year	Neither winter nor nocturnal snow very great

3. Principal Highlands

Highlands are found in all continents. Some of the best known highland regions are situated in the tropics and northern hemisphere. The tropical highlands support many large cities (Chapter XII). The most continuous and perhaps the largest highland areas fill much of South America. We have in Equatorial East Africa the Abyssinian Highlands, Mt. Kilimanjaro (6010 m) Mt. Meru (4730 m), Mt. Kenya (5600 m) the mountainous area of Leikipea (4270 m), Mt. Elgon (4230 m) and the Ruwenzori Range (5500). These highlands are related to the Rift Valley Fracture System, stretching nearly 6400 km from the Jordan Valley to south of the Zambessi. In tropical West Africa is the volcano Mt. Cameroun (4075 m). The mountain chains of Java, Sumatra, Borneo, New Guinea and Sulevesi (Celebes) are other tropical highlands. The tropical parts of the Andes belong to the Western Highlands of South America. The western Highlands extend 7000 km from Panama to Cape Horn. The Andes in Colombia are well defined ranges, with deep valleys, converging near Pasto. South of Ecuador, the Andes have three ranges across Peru and converge to Cerro de Pasco in Peru. To the south, there are two ranges, one on either side of the Bolivian Plateau, uniting at Aconcagua (7034 m) in Agentina, beyond which a single range extends to Cape Horn. Mt. Roraima of the Sierra Pacaraima, the Serra Roncador and the Brazilian Highlands are other tropical highlands of South America. The Andes constitute the mountain system of the entire length of South America, narrowest in the south and broadest in the Central and Bolivian Andes and in the north. The Andes are beyond comparison the longest unbroken mountain range in the world. They are are also among the highest mountain ranges, second only to the Pamir–Karakoram–Himalaya mass. The central Altiplano of northern Chile and Bolivia (Plates 1-1, 9-1) is an extensive, barren, broad, plateau-like valley, 35-100 km wide and 2450–3950 m above mean sea level, separated by the two parallel chains of mountains of the Andes. The Bolivian Altiplano is the broadest and highest. The valley in between the ranges of South Chilean Andes is a high level plateau at 3050–3960 m. The Puna Basin lies at 3300–3960 m.

In the north temperate zone are the great Alpine-Himalayan Systems in the Old world. The Himalaya forms an integral part of high Asia and is *par excellence* the mountain of India (Plate 1-1). The mountains of India fall under two broad groups, viz. the Peninsular and the Extra-Peninsular. The Peninsular group embraces the Western Ghats, the Eastern Ghats, the Aravalli, Vindhya and Satpura Ranges, which are relicts of an ancient high plateau. The Himalaya belongs to the Extra-Peninsular Group. The Himalayan System embraces a series of parallel ranges, extending east-west for almost 3000 km, and varying in width from 80 km to 300 km. This system is connected with the Hindu Kush, Kuen Lun, Tien Shan in the Pamir Knot. The elevated areas are continued westwards across the Caucasus and Armenian Highlands to the Alpine system. The Tibetan Plateau lies between the Kuen Lun and the Himalaya. Strictly speaking, Tibet is not a tableland but a rugged mountainous country at a mean elevation of 47500–4880 m (Plate 9-2). The Pamirs are a congregation of chains of mountains stretching east-west; they are high steppes without outlet, with the valleys at mean elevation of 3660 m and mountains 5180 m.

The principal elevated areas of Europe are the Central Uplands and Plateau, the northwestern highlands and the southern mountain ranges. The Central Uplands and Plateau comprise a continuous highland of hilly area, from south central France to north Czechoslovakia, 150–160 m above mean sea level. The Northwestern Highlands comprise Scandinavia and are higher than the Central Uplands. The southern mountains comprise the Alpine System of mean elevation of 3000 m. It is nearly 755 km long on the inner arch and 1300 km on the outer, with a width of about 150–250 km. The Alps comprise numerous ranges, with longitudinal valleys and the whole forming a broad mass convex in the north. The other important mountains of the North Temperate of the Old World are the Altai, Pyrenees, Sierra Nevada of Spain, the Apenines and the Caucasus.

The mountains of the North Temperate of the New World embrace the principal mountains of Canada, the United States and Mexico. In Canada, the elevated areas are the Eastern or Laurentian Highlands, the Appalachian Highlands and the Western Highlands or the Cordilleran mountain system. The Appalachian and the Cordilleran systems continue southward into the United States. In marked contrast to the Alpine-Himalayan Systems, the New World mountains are mainly meridional in trend. The Cordillera system extends, with a width of 640–1600 km from the extreme northwest corner of Alaska to the Isthmus of Panama, thence to merge with northern-most spurs of the Andes of South America. The whole highland is almost 16,000 km long and almost half the world across. The highest

Plate 1-1. 1. The Himalaya—a highland of rugged terrain. 2. The Bolivian Altiplano.

mountains in the Cordilleran System are in Alaska and are true arctic mountains.

The Canadian Cordillera comprise two nearly parallel system, separated by lower mountains and plateaux. The Rocky Mountains form the eastern ranges. In the United States we have a complex of many mountain chains and ranges, often widely separated by wide rolling high plateaux. The Cordilleran plateaux include the Mexican Plateau and the Utah Plateau.

The principal elevated areas of the southern hemisphere are the mountains of Australia, New Zealand, the Chilean Andes, (South of Bolivia) the Patagonian Steppes, the Andes of Tierra del Fuego and the mountains of Antarctica. Some of the socalled mountains of Australia are really extensive plateaux, intersected by valleys.

4. References

1. Heath, D. & D. R. Williams. 1977. Man at High Altitude: A pathophysiology of acclimatization and adaptation. Edinburgh: London: New York. Churchill Livingstone.
2. Hopkins, A. D. 1921. Bioclimatic zones of the Continents, with the proposed designations and classifications. *J. Washington Acad. Sci.*, 11: 227–229.
3. Mani, M. S. 1968. Ecology and Biogeography of High Altitude Insects. The Hague: Dr W. Junk-BV Publishers. Series Entomologica Vol. 4 pp. xvi+528, figs, 80.
4. Mani, M. S. 1974. Fundamentals of High Altitude Biology. New Delhi: Oxford & IBH Publishers. pp. 196.
5. Mani, M. S. 1978. Ecology and Phytogeography of High Altitude Plants of the Northwest Himalaya: An Introduction to High Altitude Botany. New Delhi: Oxford & IBH London: Chapman & Hall. pp. 205. pls. xxiv.

The physical environment of the highlands

L. E. Giddings

From the human perspective, the highlands present a very severe environment. Most of its harshness can be traced to only two factors, atmospheric pressure and temperature. Both are much lower than in adjacent lowlands, and the pressure of the atmosphere is probably near the long-term low limit for the survival of men. Other factors also contribute to the inhospitability of the highlands, but they are mostly secondary results of the low temperatures and pressures of the air.

For the use of scientists in all fields, this chapter presents the physical properties of the highlands atmosphere as known at present. Discussions of pressure and temperature are followed by a treatment of other properties of the atmosphere. Finally, constants and formulas will help the scientist calculate many others. Representative climatological data will illustrate typical highland conditions. A discussion of the radiation environment is presented in Chapter III.

1. Atmospheric Pressure

Since a large part of the pioneering work in science was done at low altitudes, sea-level air furnished a logical reference point for pressure. The standard for chemists, 760 torr (mm of mercury), and the meteorological standard of 1013.25 millibars, represent the average sea level pressures for mid-latitudes. However, even at sea level, variations in weather normally cause a variation as great as five or ten per cent. For precise tasks, such as the laboratory determination of vapour densities, deviations from the standard must be taken into consideration.

Pressure decreases with altitude. In simplest terms, this occurs because the pressure at a given place measures the weight of air directly above it. By using this concept, approximate values of the pressure profile can be calculated, as shown below.

Consider an infinitesimal column of air of unit cross section, with pressure P at its base and pressure $P - dP$ at its top. Equation 2.1 shows that the pressure drop, $-dP$, is given by the weight of the column:

$$-dP = \rho g dh. \tag{2.1}*$$

First, let's assume that air in our atmosphere behaves like a liquid, a fluid with a constant density. For this case, integration of 2.1 would give:

$$P_h - P_0 = -\rho g h$$

* Symbols are identified in Table 2-7. Only clarifications are given in the text.

where h is the height above sea level and P_0 is the pressure at sea level. Such an atmosphere would have an upper boundary, much like the surface of the oceans.

At such an upper surface, the pressure would be zero, since there is only a vacuum above it. Its altitude, H, would therefore be

$$H = \frac{P_0}{\rho g}.$$

Substituting the equation of pressure from the ideal gas law in the for $P = \rho RT/M$,

$$H = \frac{RT}{Mg}. \tag{2.2}$$

For the arbitrary case of dry air, with $T_0 = 300°K$,

$$H = \frac{RT_0}{Mg} = \frac{8.314 \times 10^7 \, \dfrac{\text{g cm}^2}{\text{sec}^2 \, °\text{K mole}} \times 300°\text{K}}{28.96 \, \dfrac{\text{g}}{\text{mole}} \times 980 \, \dfrac{\text{cm}}{\text{sec}^2} \times 10^5 \, \dfrac{\text{cm}}{\text{km}}} = 8.79 \text{ km}.$$

For this liquid-like atmosphere, pressure would decrease linearly from sea level to the top of the atmosphere at the rate of 760 mm/8.79 km = 86.4 torr per kilometre of altitude, as shown in Fig. 2-1.

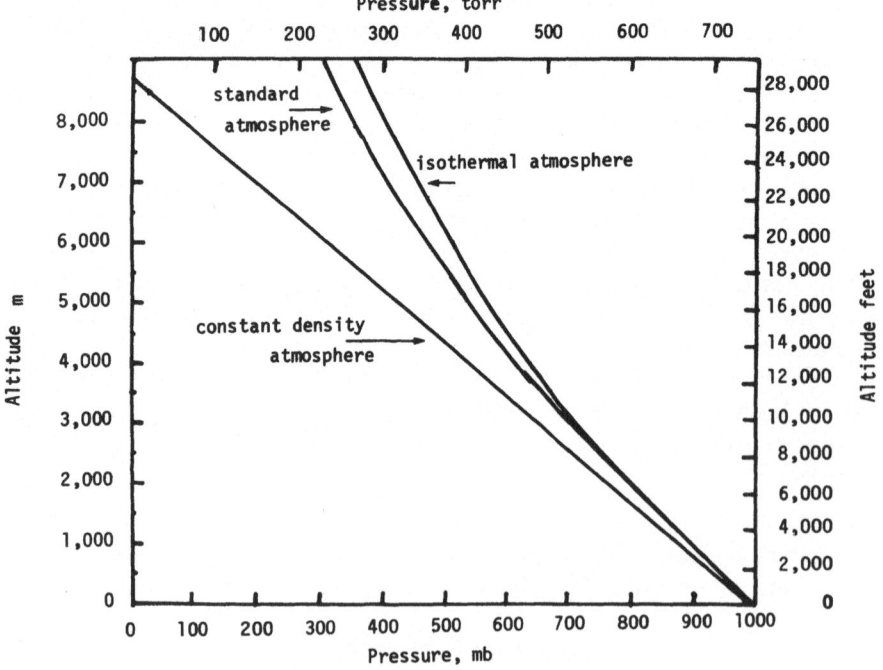

Fig. 2-1. Variation of pressure with altitude.

H has some practical uses, as will be noted later, but in our atmosphere, density does not remain constant. Instead, density decreases with altitude, to nearly zero in outer space.

We may be closer to our real case if we assume that temperature remains constant throughout the atmosphere. Inserting the perfect gas law formula for density, $\rho = PM/RT$, into equation 2.1, we obtain

$$\frac{dP}{P} = \frac{Mg}{RT} \, dh. \tag{2.3}$$

Integrating,

$$P = P_0 d^{(Mg/RT)h} = P_0 d^{h/H} \tag{2.4}$$

where H maintains its definition from 2.2.

Under this assumption, pressure will decrease exponentially with altitude. By differentiation,

$$\frac{dP}{dH} = \frac{P_0}{8.79} \exp{(h/8.79)} \bigg|_{P_0=760 \, \text{torr}} = 86.5 \, \frac{\text{torr}}{\text{km}}. \tag{2.5}$$

This sea level value of the lapse rate, about equal to 125 mb per kilometre (25 torr per thousand feet) is quite acceptable for most purposes.

Of course, our atmosphere is also not isothermal; temperature normally decreases with altitude. Still, as Fig. 2.1 shows, this second model represents it fairly well, considerably better than our first case. For more details on deriving these and other equations, the reader is referred to an excellent, lucid Russian introductory text on meteorology (8), which is available in English translation (9).

As a very practical matter, the highlands scientist needs to know the relationship of average ambient pressure to altitude. These data are presented in Table 2-1, together with various other parameters.

In addition, equations of the pressure as a function of altitude are conveniently presented as polynomials. The following are useful equations which present the very lowest and highest pressures that can be expected at a given altitude in the absence of disastrous weather phenomena, such as hurricanes. Equation (2.7) represents the standard mid-latitude atmosphere to about 1 torr. Values of h must be given in metres.

Low pressure extreme:
$$P = 732.6 - 0.0967 \times h + 0.00000421 \times h^2 \tag{2.6}$$

Standard mid-latitude pressure:
$$P = 759.4 - 0.0882 \times h + 0.00000346 \times h^2 \tag{2.7}$$

High pressure extreme:
$$P = 788.6 - 0.0869 \times h + 0.00000353 \times h^2. \tag{2.8}$$

These equations were taken from an article whose primary emphasis was chemical instead of meteorological (3). Much more precise equations can be

Table 2-1. Standard atmospheres.

15° North, Annual

h Altitude metres	P Pressure torr	t Temperature °C	T Temperature °K	T_v Virtual temperature °K	Relative humidity %	Precipitable water cm
0	760.0	26.5	299.65	302.59	75	4.07
1000	678.0	20.5	293.65	295.89	75	2.43
2000	603.3	14.5	287.65	289.34	75	1.28
2250	585.7	13	286.15	287.72	75	1.06
2500	568.5	13.8	286.95	287.74	35	0.90
3000	535.6	10.4*				
3500	504.2	7.1*				
4000	474.3	3.75	276.90	277.36	35	0.42
5000	418.6	-2.9*				
6000	368.4	-9.65	263.50	263.71	35	0.13
8000	282.3	-23.05	250.10	250.17	30	0.03

30° North, January

h Altitude metres	P Pressure torr	t Temperature °C	T Temperature °K	T_v Virtual temperature °K	Relative humidity %	Precipitable water cm
0	765.8	14	287.15	288.52	80	2.27
1000	679.9	11	284.15	285.24	70	1.44
2000	602.7	8	281.15	281.86	50	0.79
2500	567.0	4.8*				
3000	533.1	1.5	274.65	275.10	45	0.38
3500	500.8	-1.7*				
4000	470.1	-5	268.15	268.39	35	0.21
5000	413.3	-11.5*				
6000	362.1	-18	255.15	255.24	30	0.06
8000	275.1	-31	242.15	242.18	30	0.01

30° North, July

h Altitude metres	P Pressure torr	t Temperature °C	T Temperature °K	T_v Virtual temperature °K	Relative humidity %	Precipitable water cm
0	760.2	28	301.15	304.58	80	4.38
1000	678.4	20.5	293.65	295.58	65	2.69
2000	603.6	15	288.15	289.54	60	1.71
2500	568.8	12.3*				
3000	535.8	9.5	282.65	283.72	60	1.05
3500	504.3	6.8*				
4000	474.4	4	277.15	277.82	50	0.62
5000	419.0	1.5*				
6000	369.0	-7	266.15	266.44	40	0.19
8000	283.6	-21	252.15	252.27	40	0.18

45° North, January

Altitude	Pressure	Temp				
0	763.6	-1	272.15	272.59	77	0.85
1000	673.0	-4.5	268.65	269.00	70	0.56
2000	592.2	-8	265.15	265.43	65	0.35
2500	555.3	-9.75*				
3000	520.3	-11.5	261.65	261.85	55	0.20
3500	487.2	-14.5*				
4000	456.0	-17.5	255.65	255.77	50	0.11
5000	398.3	-23.5*				
6000	346.8	-29.5	243.65	243.70	45	0.03
8000	260.1	-41.5	231.65	231.66	35	0.01

60° North, January

Altitude	Pressure	Temp				
0	760.2	-16	257.15	257.28	80	0.42
1000	666.0	-14	259.15	259.31	70	0.30
2000	583.3	-17.2	255.95	256.09	70	0.19
2500	545.6	-18.8*				
3000	510.0	-20.4	252.75	252.86	65	0.11
3500	476.6	-22	251.15	251.24	60	0.08
4000	445.1	-25.4	247.75	247.82	60	0.06
5000	387.0	-32.2	240.95			
6000	335.2	-39	234.15	234.17	50	0.01
8000	248.1	-52.6	220.55	220.55	40	

45° North, July

Altitude	Pressure	Temp				
0	760.2	21	294.15	296.22	75	2.97
1000	676.7	16.5	289.65	291.14	65	1.82
2000	601.2	12	285.12	286.19	55	1.06
2500	566.1	9*				
3000	532.8	6	279.15	279.78	45	0.61
3500	501.0	3*				
4000	470.9	0	273.15	273.55	40	0.35
5000	415.0	-6*				
6000	363.8	-12	261.15	261.30	30	0.10
8000	278.9	-25	248.15	248.21	30	0.03

60° North, July

Altitude	Pressure	Temp				
0	757.6	14	287.15	288.45	75	2.01
1000	672.1	8.6	281.75	282.68	70	1.25
2000	594.9	3.2	276.35	277.06	70	0.74
2500	559.2	0.5*				
3000	525.2	-2.2	270.95	271.45	65	0.40
3500	493.0	-4.9	268.25			0.33
4000	462.5	-7.6	265.55	265.89	60	0.29
5000	406.2	-13	260.15	260.38	55	0.15
6000	355.6	-20	253.15	253.28	50	0.07
8000	269.4	-34	239.15	239.18	40	0.01

* Interpolated values

Table 2-1 (Continued)

75° North, January

h Altitude metres	P Pressure torr	t Temperature °C	T Temperature °K	T_v Virtual temperature °K	Relative humidity %	Precipitable water cm
0	760.2	−24	249.15	249.22	80	0.21
1000	663.4	−21	252.15	252.23	65	0.15
1500	620.1	−19.5	253.65	253.74	60	0.12
2000	579.4	−22.25	250.90	250.98	60	0.09
2500	541.1	−25	248.15			
3000	504.9	−27.75	245.40	245.45	55	0.05
3500	470.8	−30.5*				
4000	438.6	−33.25	239.90	239.93	50	0.03
5000	379.8	−38.75*				
6000	327.7	−44.25	228.90	228.91	45	0.01
8000	241.4	−55.25	217.90	217.90	40	

75° North, July

h Altitude metres	P Pressure torr	t Temperature °C	T Temperature °K	T_v Virtual temperature °K	Relative humidity %	Precipitable water cm
0	759.4	5	278.15	278.92	85	1.19
1000	671.5	2.4	275.55	276.19	75	0.69
1500	631.1	1.1	274.25			0.58
2000	593.0	−0.2	272.95	273.46	65	0.49
2500	557.0	−1.5	271.65	272.14	65	0.32
3000	522.9	−4.75	268.40			0.25
3500	490.5	−8*				
4000	459.8	−11.25	261.90	262.13	55	0.19
5000	402.9	−17.75*				
6000	351.9	−24.25	248.90	248.98	45	0.05
8000	265.4	−37.25	235.92	35	0.01	

* Interpolated values

found in the Standard Atmospheres (2, 6). These extremes represented by equations (2.6) and (2.8) may serve the purposes of chemists. Much more authoritative information on climatic extremes are available in NASA documents concerning launching conditions for satellites (1).

It is important to note that these equations yield pressures in units of torr (millimetres of mercury). Pressures will be specified in similar units throughout this book, although other prameters will be expressed in the metric system. Tables for conversion of pressures to other units can be found in the chapter on barometry.

2. Air Temperature

Air temperatures at the surface of the earth are lower in the highlands than in the plains, but it is not easy to derive a useful lapse rate from theory. One must account for time of day, time of year, latitude, water vapor content of the air, surface morphology, and many other factors in an acceptable formulation of surface temperature and its lapse rate.

Empirically there is in free air usually a decrease of about 0.6° for each 100 m increase in altitude in middle latitudes (3.3°F for each 1000 feet). Average values range from 0.3° to 0.75° in different seasons at different latitudes, and extremes are very varied. In areas with inversions, the temperature increases with altitude, and inversions are not at all uncommon.

Even more complex is surface air temperature. At the surface, many factors must be taken into consideration. Bare soil, short grass, trees, and water surfaces all have different influences on the local surface temperature. Still, it can be shown that the lapse rate for surface temperatures is about the same as the lapse rate for upper air at the same altitude.

3. Standard Atmospheres

For the practical needs of civil aviation in its early years, standard tables were constructed to show the variation of pressure with altitude. With these tables altimeters could be constructed with some degree of uniformity. In recent years these tables, now called standard atmospheres, have been extended to include many other properties of the atmosphere. They are now more useful for rocketry and satellite calculations than for aviation.

At this time the primary sources of high altitude parameters are the ICAO Standard Atmosphere (6) and the supplements to the U.S. Standard Atmosphere (2). These two references contain a wealth of practical data of use to the highlands scientist, even though they were designed for upper air use, not for ground level in the highlands.

Table 2-1 presents some temperatures, pressures and other data from the supplements, as well as some other useful information. Strictly speaking, all values in these tables refer only to the northern hemisphere, and similar atmospheres have not yet been formulated for the southern hemisphere. We can assume that they will be useful at least as a first approximation to conditions in the southern hemisphere.

The temperatures to be found in the main tables of these standard atmospheres are the so-called *virtual temperatures*, not actual temperatures. A virtual temperature 'is the fictitious temperature which dry air must have at the given pressure P, in order to have the same density ρ, as a water vapour-air mixture at the same pressure P, temperature T, and vapour pressure e (that is, the partial pressure of water vapor in the air). The assumption that the mixture behaves as a perfect gas eliminates the necessity for considering minor deviations from the perfect gas law such as for the compressibility factor of air which is a function of pressure, temperature and relative humidity. The error in computed densities resulting from the assumption that air is a perfect gas may approach 0.05 per cent below 10 km (2).

A virtual temperature used with the molecular weight of dry air, using the perfect gas law, will yield quite good values of pressure and density. Since the ICAS Standard Atmosphere (6) (equal to the mid-latitude spring/fall atmosphere of the supplements (2)) is dry, its virtual and actual tempeatures are equal.

Actual temperatures and relative humidities are the defining variables for each altitude. Given the variation of temperature with altitude, and assuming the ideal gas law, pressures can be calculated as a function of altitude for dry air. Moisture must be handled in some way to allow the same calculation for moist air.

Further details on the origin and use of the standard atmospheres are available in the detailed introductions to both volumes.

4. Composition of the Atmosphere

The air is a mechanical mixture, mainly of five gases – nitrogen, oxygen, argon, carbon dioxide and water vapour making up 99.997% by volume below 90 km. Recent explorations by rockets show that, at least up to 50 km, the component gases are fairly constant in proportion. It is highly compressible such that its lower layers are very much more dense than above. The mean density decreases from about 1.20 kg/m^3 at the surface of the Earth to about 0.70 kg/m³ at an elevation of 5000 m. Water vapour amounts to 4% by volume at sea level, but above 10–12 km elevation it is almost absent. Ozone is concentrated mainly at elevations between 15 and 35 km. Variations are observed specially in the proportions of water vapour and ozone with seasons and latitude; the ozone content is low over the equator and high over latitudes to the north of 50°NL especially in spring. Water vapour content is closely related to temperature and is greater in summer and low latitudes. The carbon dioxide content is about 315 parts on an average per million and varies in higher north latitudes. At 10°NL it ranges from 310 parts in late summer to 318 parts per million in spring. The average molecular weight of air scarcely varies with altitude below 50 km if

18

Table 2-2. The composition of the Atmosphere.

Constituent gas and formula	Content, percent by volume	Content variable relative to its normal	Molecular weight*
Nitrogen (N₂)	78.084	—	28.0134
Oxygen (O₃)	20.9476	—	31.9988
Argon (Ar)	0.934	—	39.948
Carbon dioxide (CO₂)	0.0314	†	44.00995
Neon (Ne)	0.001818	—	20.183
Helium (He)	0.000524	—	4.0026
Krypton (Kr)	0.000114	—	83.80
Xenon (Xe)	0.0000087	—	131.30
Hydrogen (H₂)	0.00005	?	2.01594
Methane (CII₄)	0.0002	†	16.04303
Nitrous oxide (N₂O)	0.00005	—	44.0128
Ozone (O₃)	Summer: 0 to 0.000007	†	47.9982
	Winter: 0 to 0.000002	†	47.9982
Sulfur dioxide (SO₂)	0 to 0.0001	†	64.0628
Nitrogen dioxide (NO₂)	0 to 0.000002	†	46.0055
Ammonia (NH₃)	0 to trace	†	17.03061
Carbon monoxide (CO)	0 to trace	†	28.01055
Iodine (I₂)	0 to 0.000001	†	253.8088

* On basis of carbon-12 isotope scale for which $C^u = 12$.
† The content of the gases marked with a dagger may undergo significant variations from time to time or from place to place *relative* to the normal indicated for those gases.

we do not consider its varying humidity. For this reason, Table 2-2 can be taken to represent the composition of dry air at any highlands location.

5. Precipitable Water

The standard atmospheres are partially defined in terms of the relative humidity values in the table. This implies that each atmosphere contains a certain amount of water vapour. This is normally measured as precipitable water, the height of the column if all the water above it should precipitate out as rain.

A description of the calculation of precipitable water in the standard atmospheres is beyond the scope of this chapter. Reference describes the calculation in a general way, and reference presents more detailed cases.

Jesske (5) discussed this problem and calculated the precipitable water values included with the standard atmospheres. These tables (2-1, 2-3 and 2-4) should give the reader some practical feeling of the variation of precipitable water in the highland atmosphere.

These data are presented in a different manner from the data in Jesske's original work. He lists the amount of water between pressure levels in the atmosphere. For our purposes, the total amount above a given location is more important. In preparing these tables, his data were rearranged and

Table 2-3. Atmospheric water vapour for certain extreme cases.

	Key West, Florida		Lake Charles, Louisiana		Brownsville, Texas		Boothville, Louisiana		New York City, New York	
Place	Key West, Florida		Lake Charles, Louisiana		Brownsville, Texas		Boothville, Louisiana		New York City, New York	
Date	July 17, 1972		September 5, 1972		October 20, 1972		November 10, 1972		December 10, 1972	
Time, GMT	0000 hours		0000 hrs		1200 hrs		1200 hrs		1200 hrs	
Surface temperature, °C	28.4		28.4		22.6		31.4		−7.7	
Surface dewpoint, °C	25.6		24.1		19.9		29.7		−7.7	
Comment	High moisture in lower levels (below 850 mb)		High moisture in middle levels (approximately 800 ml), surface temperature inversion		High moisture in lower levels (below 80 mb), inversion		High extreme surface dew-point, temperature		High level moisture (550 to 880 mb)	
	Pressure mb	Precipitable water cm	Pressure mb	Precipitable water cm	Pressure mb	Precipitable water cm	Pressure mb	Precipitable water cm	Pressure mb	Precipitable water cm
	1017	5.85	1017	5.30	1016	4.79	1013	4.00	1016	3.38
	1000	5.50	1000	5.06	1000	4.54	1000	3.72	1000	3.31
	850	3.06	931	3.89	850	2.39	781	0.88	912	2.73
	700	1.53	774	1.83	796	1.83	767	0.74	889	2.54
	642	1.08	718	1.30	700	0.94	747	0.64	558	0.39
	591	0.76	666	0.89	586	0.24	588	0.32	500	0.22
	500	0.39	594	0.43	500	0.15	500	0.22	458	0.14
	400	0.13	500	0.11	400	0.06	400	0.07	400	0.05
	300	0.01	400	0.04	300	0.00	372	0.04	300	0.00
	279		267		265		289		250	

	Tucson, Arizona		Caribou, Maine		Albany, New York		Ely Nevada		Rapid City, South Dakota	
Place	June 30, 1972		January 10, 1973		January 30, 1973		December 5, 1972		January 10, 1972	
Date	0000 hrs		1200 hrs		1200 hrs		1200 hrs		1200 hrs	
Time, GMT	39.4		−34.5		−16.3		−22.3		−19.5	
Surface temperature, °C	−7.6		−38.6		−27.3		−24.6		−28.5	
Surface dewpoint, °C	High extreme surface temperature		Low extreme surface temperature and dewpoint at surface and 800 mb		Low surface moisture and maximum moisture at 850 mb		Low surface moisture and maximum moisture at 700 mb, surface temperature inversion		Low moisture at 750 and 500 mb, surface temperature inversion	
Comment	Pressure mb	Precipitable water cm	Pressure mb	Precipitable water cm	Pressure mb	Precipitable water cm	Pressure mb	Precipitable water cm	Pressure mb	Precipitable water cm
	922	1.28	987	0.28	1008	0.20	798	0.17	907	0.13
	907	1.21	966	0.27	959	0.18	795	0.16	893	0.13
	850	0.91	898	0.23	915	0.15	781	0.15	864	0.11
	700	0.39	850	0.20	832	0.09	754	0.12	836	0.10
	578	0.16	804	0.17	775	0.05	690	0.05	729	0.06
	526	0.09	745	0.12	721	0.03	679	0.05	700	0.05
	500	0.07	700	0.09	665	0.02	592	0.02	670	0.04
	400	0.02	617	0.05	649	0.01	532	0.01	630	0.03
	300	0.00	530	0.01	500	0.00	500	0.01	500	0.01
	275		471		459		412		400	

Table 2-4. Atmospheric water vapor during severe storms.

Place	Rush Springs, Oklahoma		Fort Still, Oklahoma		Hinton, Oklahoma		Chickasha, Oklahoma	
Date	May 15, 1968		May 15, 1968		May 15, 1968		June 8, 1968	
Time, GWT	1701 hrs		1801 hrs		1900 hrs		1648 hrs	
Surface temperature, °C	25.8		28.8		27.4		28.2	
Surface dewpoint, °C	22.3		15.5		13.4		15.2	
Comment	Before dry line passage		During dry line passage		After dry line passage		Ahead of front	
	Pressure mb	Precipitable water cm	Pressure mb	Precipitable water cm	Pressure mb	Precipitable water cm	Pressure mb	Precipitable water cm
	961	3.03	966	1.90	949	2.26	968	2.81
	879	1.65	910	1.29	907	1.85	938	2.42
	850	1.36	877	0.96	892	1.72	891	1.82
	808	1.20	850	0.71	850	1.43	850	1.34
	760	0.78	700	0.49	715	0.72	792	0.80
	718	0.53	666	0.42	647	0.49	700	0.38
	700	0.47	594	0.29	600	0.35	600	0.17
	600	0.27	500	0.11	500	0.11	500	0.06
	500	0.11	457	0.05	462	0.05	464	0.03
	400		400		400		400	

	Cordell, Oklahoma		Tinker AFB, Oklahoma		Pauls Valley, Oklahoma		Sheppard AFB, Texas	
Place	Cordell, Oklahoma		Tinker AFB, Oklahoma		Pauls Valley, Oklahoma		Sheppard AFB, Texas	
Date	June 8, 1966		June 5, 1966		June 9, 1966		May 24, 1966	
Time, GMT	2000 hrs		2000 hrs		2000 hrs		2000 hrs	
Surface temperature, °C	23.0		30.2		17.4		24.4	
Surface dewpoint, °C	2.0		11.2		7.4		4.4	
Comment	Behind front		Before front		Behind front		Behind front (overrunning)	
	Pressure mb	Precipitable water cm	Pressure mb	Precipitable water cm	Pressure mb	Precipitable water cm	Pressure mb	Precipitable water cm
	951	1.86	963	2.04	984	3.20	983	1.77
	901	1.57	915	1.55	963	3.05	967	1.69
	850	1.26	850	0.82	910	2.67	878	1.43
	803	0.95	818	0.47	850	2.01	850	1.37
	751	0.63	802	0.32	802	1.58	800	1.18
	700	0.33	700	0.15	700	0.88	735	0.90
	622	0.16	600	0.07	595	0.29	700	0.69
	549	0.08	540	0.04	554	0.18	595	0.26
	500	0.05	500	0.02	500	0.07	500	0.07
	400		400		400		400	

summed in a way that would present the total water vapour in the column of air above a given location.

The data in Table 2-1 present the water in the atmosphere from the indicated altitude of up to 10,000 m. Tables 2-3 and 2-4 present the same data up to the lowest pressure level reading of the radiosonde, which is normally 400 mb (the lowest pressure level is given in the tables). In practice, since the upper atmosphere contains very little water, these numbers can be used for the total column of water. In no case should they be more than 0.1 cm low, and in most cases they will be within 0.02 or 0.03 cm.

For the practical problem of estimating precipitable water above a given highlands location, one might well assume, as a first approximation, the value from the most appropriate of the standard atmospheres in Table 2-1. He should recognize from Tables 2-3 and 2-4 the possible errors in his assumed value. In addition, he should also know that the interaction of the surface with water vapour is largely unknown. If he needs precise values, he must measure them.

6. Regimes of Temperature and Pressure

Actual surface air temperature and precipitation vary with season of the year, geography, location with respect to the global currents, etc. Although these subjects are beyond the scope of this works, Figs. 2-2 to 2-12 present the actual measurements for a number of highlands location (4).

7. Other Properties

Table 2-5 presents many useful properties of air as a function of altitude. These data have been extracted from the volume describing the U.S. Standard Atmosphere (coincident with the ICAO atmosphere) (6). The variation of many properties appears to be greater than appreciated by many highlands scientists.

Since Table 2-5 refers only to a few specific altitudes in a single standard atmosphere, Table 2-6 presents formulas for the same properties. The reader may calculate the most representative values for his location by using measured values of temperature and pressure or by reference to the most appropriate entries in Table 2-1. For the convenience of the reader, Table 2-6 also includes values of the physical constants needed in these calculations. Table 2-7 includes symbols and units, including a few from other chapters.

8. Upper Air

This book is properly concerned only with the surface of the earth, which extends only as high as about 9000 m above sea level. However, for

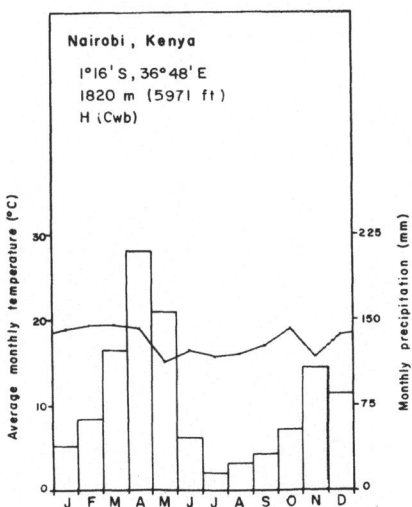

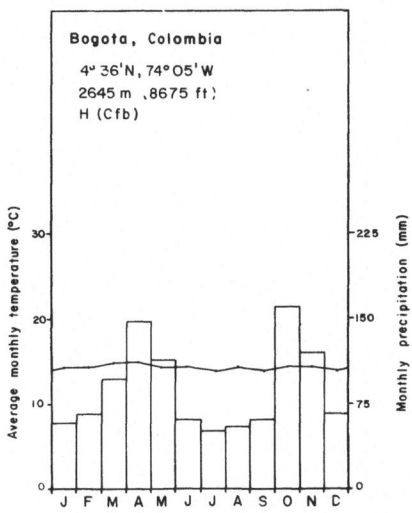

Fig. 2-2. Temperature and precipitation in high areas. Bars represent precipitation, and lines represent temperatures. Graphs are presented in order of latitude without attention to hemisphere.

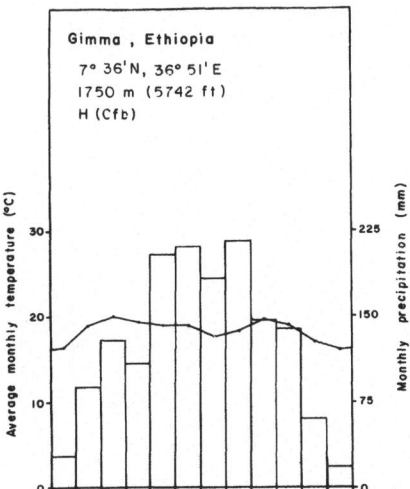

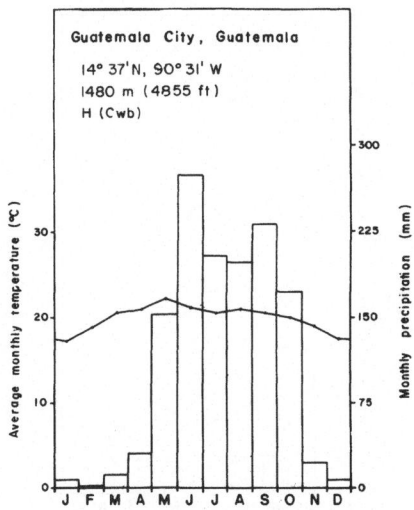

Fig. 2-3. Temperature and precipitation in high areas. Bars represent precipitation, and lines represent temperatures. Graphs are presented in order of latitude without attention to hemisphere.

25

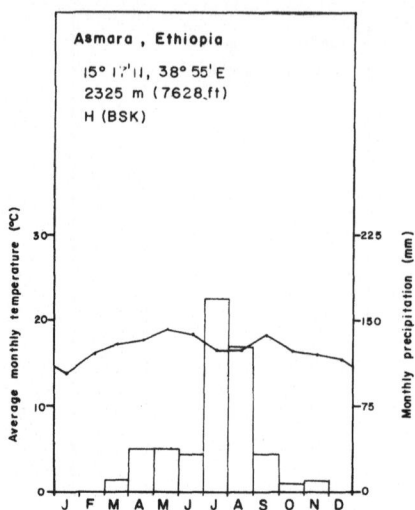

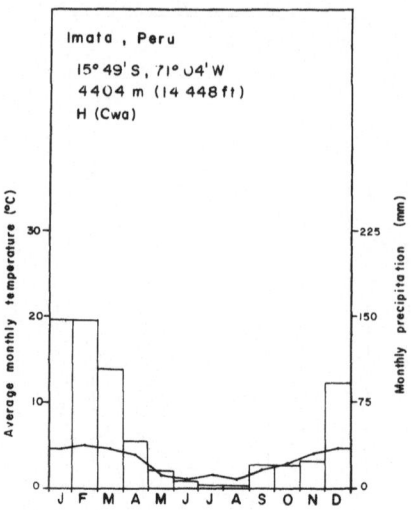

Fig. 2-4. Temperature and precipitation in high areas. Bars represent precipitation, and lines represent temperatures. Graphs are presented in order of latitude without attention to hemisphere.

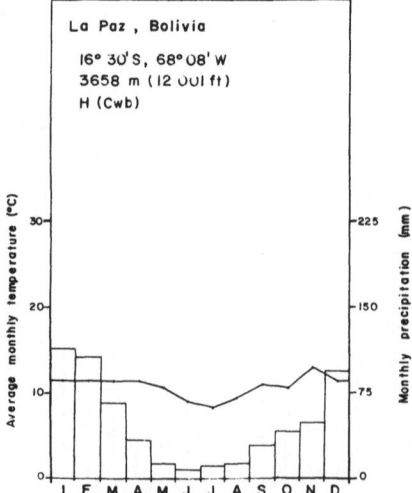

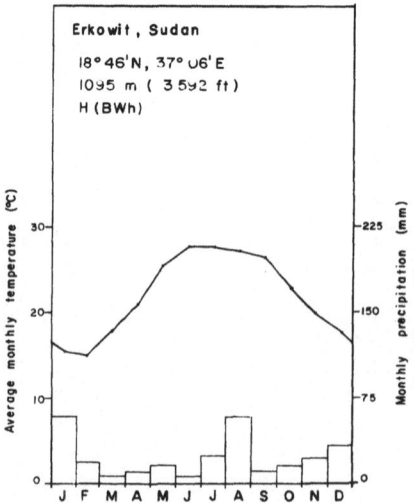

Fig. 2-5. Temperature and precipitation in high areas. Bars represent precipitation, and lines represent temperatures. Graphs are presented in order of latitude without attention to hemisphere.

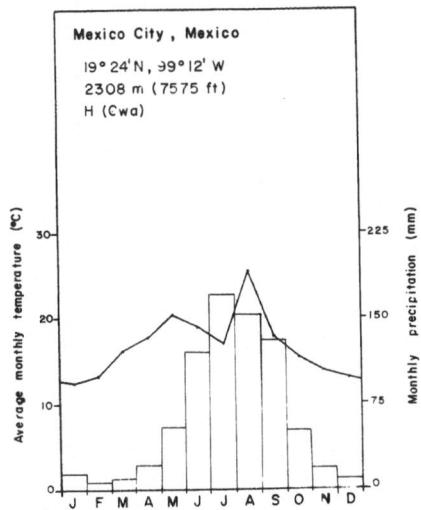

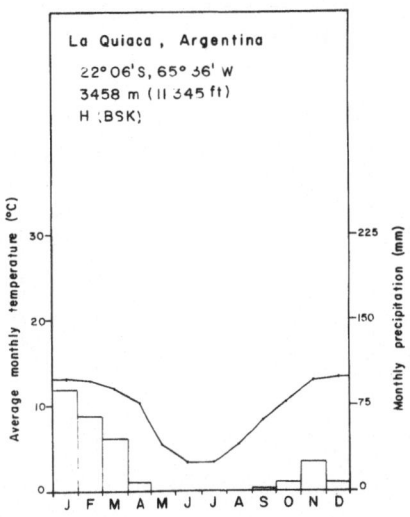

Fig. 2-6. Temperature and precipitation in high areas. Bars represent precipitation, and lines represent temperatures. Graphs are presented in order of latitude without attention to hemisphere.

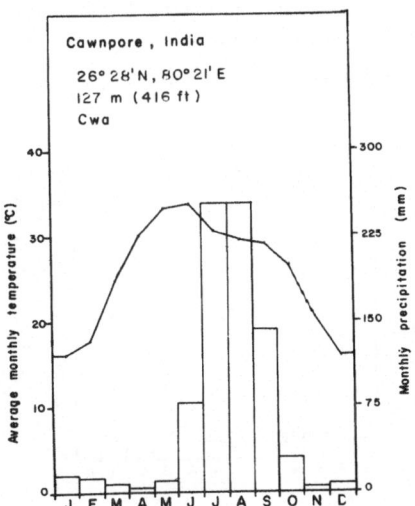

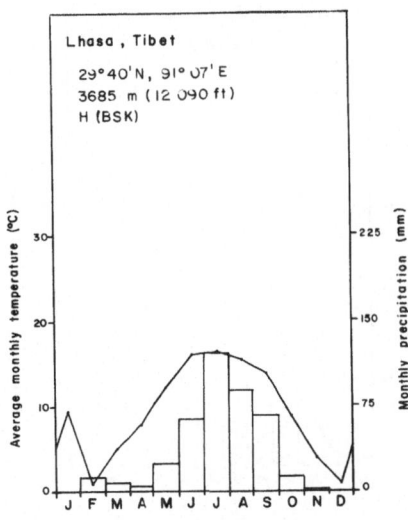

Fig. 2-7. Temperature and precipitation in high areas. Bars represent precipitation, and lines represent temperatures. Graphs are presented in order of latitude without attention to hemisphere. Cawnpore in the plains of India is included by way of comparison.

27

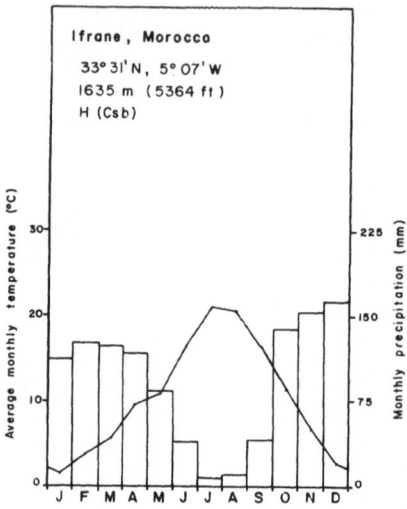

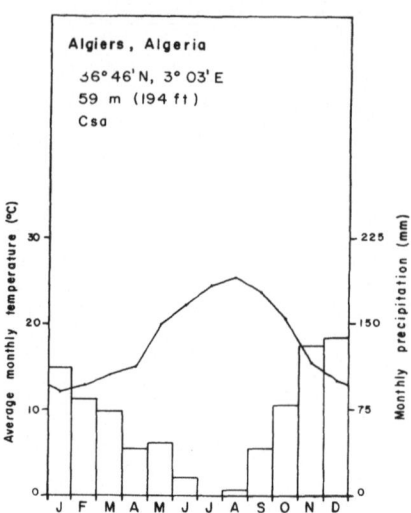

Fig. 2-8. Temperature and precipitation in high areas. Bars represent precipitation, and lines represent temperatures. Graphs are presented in order of latitude without attention to hemisphere. Algiers is in the plains and is included for comparison.

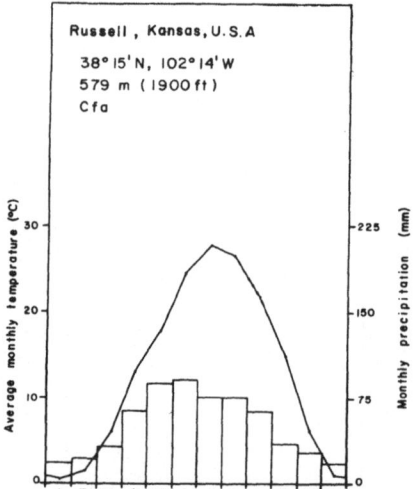

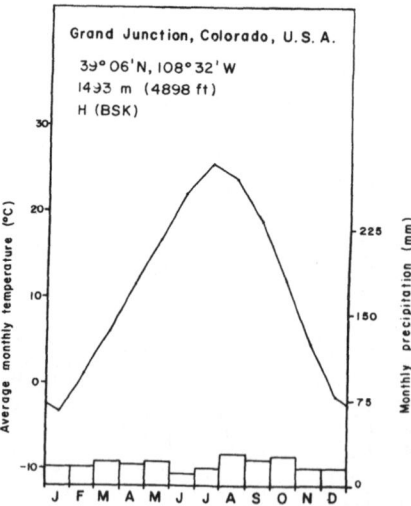

Fig. 2-9. Temperature and precipitation in high areas. Bars represent precipitation, and lines represent temperatures. Graphs are presented in order of latitude without attention to hemisphere.

28

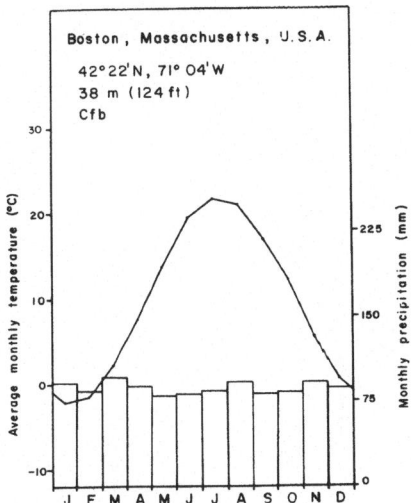

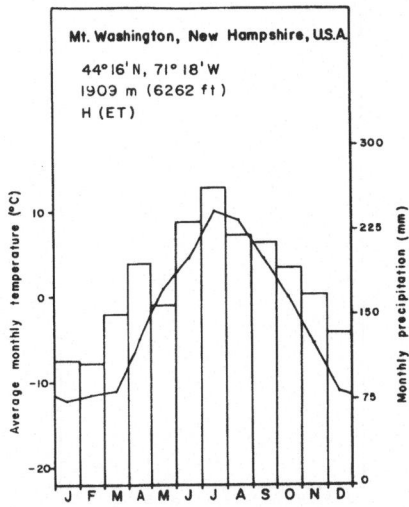

Fig. 2-10. Temperature and precipitation in high areas. Bars represent precipitation, and lines represent temperatures. Graphs are presented in order of latitude without attention to hemisphere. Boston at sea level is included by way of comparison.

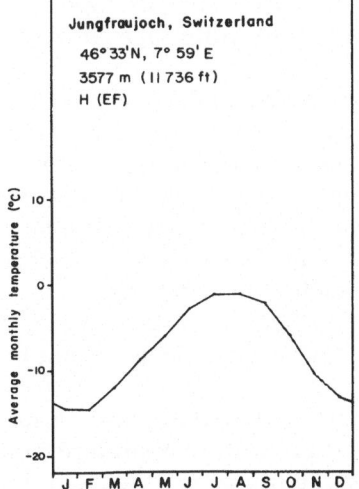

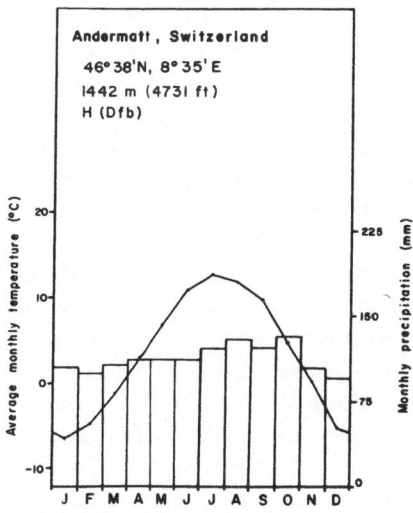

Fig. 2-11. Temperature and precipitation in high areas. Bars represent precipitation, and lines represent temperatures. Graphs are presented in order of latitude without attention to hemisphere.

29

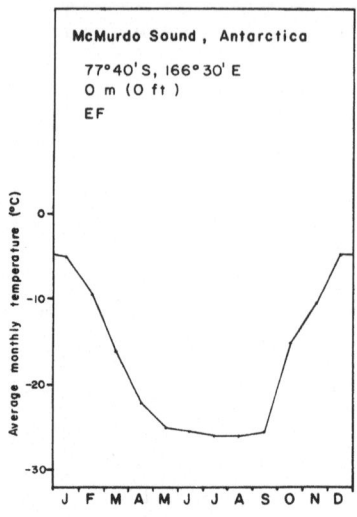
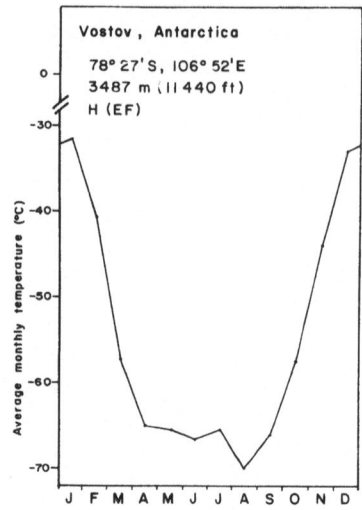

Fig. 2-12. Temperature and precipitation in high areas. Bars represent precipitation, and lines represent temperatures. Graphs are presented in order of latitude without attention to hemisphere. McMurdo Sound at sea level is included for comparison.

perspective, it is interesting to note the variation of properties in upper air. Figure 2-13 illustrates several important properties to quite high altitudes (7).

Above 50 km the components of air begin to ionize under the effect of short wavelength solar radiation. As a result, the average molecular weight decreases quite strongly until it almost coincides with the average atomic weight of air. Still higher, fractionating of atomic weights reduces it even further as light species, such as hydrogen, begin to predominate. Pressure continues to decrease with altitude as expected, and density finally reaches extremely low values. A very high vacuum is reached in outer space.

Although temperature decreases with altitude in the lower atmosphere, it begins to rise at about 18 km and fall again at about 50 km. Texts should be consulted for further details and explanations.

Finally, the value of $H = RT/Mg$, which was calculated before, also varies. This useful quantity, called the scale height, changes with temperature and mean molecular weight. Its use for measuring altitude is discussed by Tverskoi (8, 9).

Further details on these and other properties of the atmosphere can be found in many texts and references. Especially valuable for orientation is the Handbook of Geophysics and Space Environment (10).

Table 2-5. Some secondary properties of the U.S. standard atmosphere.

h geopotential altitude metres	t, temperature °C	P, Pressure torr	ρ, Density kg m^{-3}	ω, Specific weight kg m^{-2} sec^{-2}	n, Number density m^{-3} ×10^{25}	$\bar{V}$ mean particle speed m sec^{-1}	L, Mean free path m ×10^{-8}	Collision frequency sec^{-1} ×10^{9}	C_x speed of sound m sec^{-1}	μ, Coefficient of viscosity kg m^{-1} sec^{-1} ×10^{-5}	η, Kinematic viscosity m^2 sec^{-1} ×10^{-5}	k, Coefficient of thermal conductivity kcal m^{-1} sec^{-1} ×10^{-6}	g, Acceleration due to gravity m sec^{-2}
9000	−43.5	230.587	0.46635	4.5604	0.96968	409.72	17.423	2.3516	303.793	1.4922	3.1997	4.9218	9.7789
8000	−35	267−020	0.52517	5.1372	1.0920	415.48	15.472	2.6854	308.063	1.5268	2.9072	5.0507	5.0507
7000	−30.5	307.981	0.58950	5.7683	1.2257	421.15	13.783	3.0556	312.273	1.5610	2.6479	5.1788	9.7851
6000	−24	353.886	0.65970	6.4572	1.3717	426.76	12.317	3.4649	316.428	1.5947	2.4174	5.3061	9.7881
5000	−17.5	405.182	0.73612	7.2075	1.5306	432.29	11.038	3.9164	320.529	1.6281	2.2118	5.4326	9.7912
4000	−11	462.339	0.81913	8.0228	1.7032	437.75	9.9193	4.4131	324.579	1.6611	2.0279	5.5582	9.7943
3000	−4.5	525.857	0.90912	8.9070	1.8903	443.14	8.9374	4.9583	328.578	1.6937	1.8630	5.6831	9.7974
2000	2	596.263	1.0065	9.8641	2.0928	448.47	8.0728	5.5554	332.529	1.7260	1.7148	5.8072	9.8005
1000	8.5	674.114	1.1116	10.898	2.3114	453.74	7.3092	6.2078	336.434	1.7578	1.5813	5.9305	9.8036
Sea Level	15	760.00	1.2250	12.013	2.5471	458.94	6.6328	6.9193	340.294	1.7894	1.4607	6.0530	9.8066

Table 2-6. Formulas and constants for calculating secondary properties of the atmosphere.

Density: $\rho = MP/RT$
Specific Weight: $\omega = \rho g$
Number density: $n = NP \times 10^6/RT$
Mean Particle Speed: $\bar{V} = \sqrt{8RT/\pi M}$
Mean Free Path: $L = RT/\sqrt{2}\,\pi\sigma^2 p \times 10^6$
Collision Frequency: $\nu = \bar{V}/L$
Speed of Sound: $C_s = \sqrt{1.40RT/M \times 10^4}$
Coefficient of Viscosity: $\mu = 1.458 \times 10^{-6} \times T^{1.5}/(T + 110.4)$
Kinematic Viscosity: $\nu = \mu/\rho$
Coefficient of Thermal Conductivity: $k = 6.324 \times 10^{-7} \times T^{1.5}/(T + 2454)$
Acceleration due to Gravity: $g_\phi = 9.780356(1 + 0.0052885 \sin^2 \phi.$
$\qquad\qquad\qquad\qquad -0.0000059 \sin^2 2\phi) \text{ m sec}^{-2}$
Absolute Temperature: $T = 273.16 + t$
  $M = 28.964$ grams/mole

  $g = 9.80665$ m/sec^2 at sea level at 45° North Latitude
  $R = 8.31432 \times 10^7$ er/°K mole
    $= 1.987$ cal/mole °K
    $= 0.082056\ 1$ atm/mole °K
  $N = 6.023 \times 10^{23}$/mole
  $\sigma = 3.65 \times 10^{-10}$ m (mean collision diameter)

Table 2-7. Symbols and units for Chapter 2 through Chapter 6.

C_s	speed of sound	m/sec
D_λ	fraction of solar constant	see Table 3-1
E	Arrhenius activation energy	kcal/mole
g	gravitation constant	m/sec^2
h	height above sea level	m
k	constant of one sort or another	
k	coefficient of thermal conductivity	kcal/m sec °K
k	rate constant	see Chapter 5
L	mean free path	m
L	latent heat of evaporation	kcal/mole
M	molecular weight	grams/mole
n	number density	m^{-3}
N	Avogadro's number	mole^{-1}
P	pressure	torr (mm of Hg) or mb
P_0	pressure at sea level	torr (mm of Hg) or mb
P	solar spectral irradiance	mW/cm^2 μm
R	ideal gas constant	see Table 3-6
t	temperature	°C
T	absolute temperature	°K
T_0	absolute temperature at sea level	°K
$\bar{V}$	mean particle speed	m/sec
z	zenith angle	°
η	kinematic viscosity	m^2/sec
λ	wavelength	μm or Å
μ	coefficient of viscosity	kg/m sec
ν	collision frequency	sec^{-1}
ρ	density	kg/m^3
σ	average collision diameter of air	m
τ	optical thickness	see Chapter 3
τ'	external optical thickness	see Chapter 3
ω	specific weight	kg/m^2 sec^2

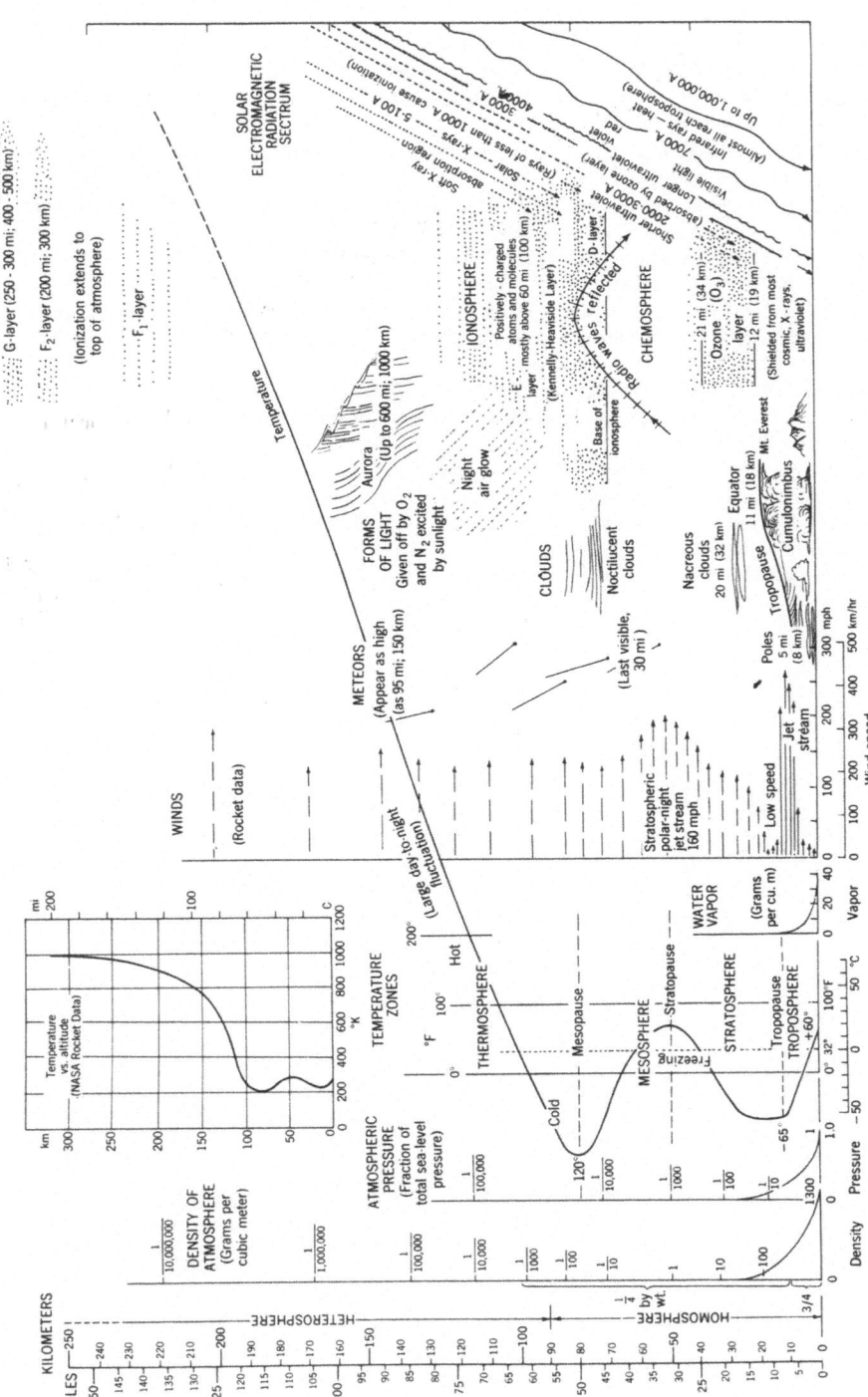

Fig. 2-13. Physical properties of the Atmosphere (from Strahler: *The Earth Sciences*, by permission).

33

9. References

1. Daniels, Glenn E. 1971. Terrestrial Environment (Climatic) Criteria guidelines for use in Space Vehicle Development 1971 Revision. NASA TM X-64587; NASA Space Vehicle Design Criteria (Environment), Surface Atmospheric Extremes (Launch and Transportation areas). NASA SP-8084 (1974).
2. ESSA, NASA, USAF US Standard Atmosphere Supplements 1966 US Government Printing Office, Washington DC.
3. Giddings, Jr. L. E. & G. Uriose. 1972. La Quimica en la altura. I– Prolongacion de ebullicion en compensacion de la altura. *Afinidad*, 29: 741–745.
4. Griffiths, J. F. 1966. Applied Climatology: An Introduction. London: Oxford.
5. Jesske, Keith W. 1974. Extreme atmosphere models, 1973. NASA TN X-58112.
6. NASA, USAF, USWB, US Standard Atmosphere, 1962. (ICAO standard atmosphere to 20 kilometres). US Government Printing Office, Washington DC. (An identical 1976 atmosphere has also been published.)
7. Strahler, G. 1972. The Earth Sciences. Harper & Row.
8. Tverskoi, P. N. 1962. Kurs Meteorologii, Fisika Atmosfery. Gidrometeorologicheskoe Izdatelstvo. Leningrad.
9. Tverskoi, P. N. 1965. Physics of the atmosphere, a course in meteorology. Israel Programme for Scientific translation. Jerusalem (N66-23462).
10. Valley, S. L. 1965. Handbook of Geophysics and Space Environment. New York: McGraw-Hill Book Co.
11. Walter J. Saucier, 1955. Principles of Meterological Analysis. Chicago: Chicago Univ. Press.

I The radiation environment

L. E. Giddings

Although often overlooked, the radiation environment is one of the most distinguishing features of the highlands. The quantity of radiation is higher than in the lowlands, and its quality differs. Its effect on the vegetation and animal life is drastic. Many of its consequences, especially in agriculture, have not yet been explored in great depth.

This chapter outlines the principal factors of the radiation environment. It deals first with the spectral distribution of the direct solar radiation, then of diffuse or indirect light, and finally its total intensity. It discusses its attenuation by haze and presents a practical way of estimating the light reaching the highlands. A discussion of the effects of the radiation environment will be postponed to a later chapter.

1. Solar Energy Outside the Atmsophere

The total radiant energy of the sun is specified by the solar constant, H_0, which is the total irradiation per unit area normal to the sun at one astronomical unit, outside the earth's atmosphere. H_0 is now taken to be 135.3 ± 2.1 milliwatts per square centimetre (1.940 ± 0.3 calories per minute-square centimetre) (17). Since the distance of the Sun from the Earth varies, the total irradiance varies from $130.9 \, \text{mW/cm}^2$ at greatest distance to $139.9 \, \text{mW/cm}^2$ at closest approach. Despite the many variations and periodicities observable in the sun, the solar constant has '... a surprisingly small, irregular variation, seldom exceeding a range of 2 per cent, with a gradual trend toward lower values. The total increase in the means of successive 5-year intervals is 0.3 per cent since 1925' (1).

The effective blackbody temperature of the sun is 5630.7°K; this is the temperature of blackbody curve normalized to the intensity of the sun (17). However, relative to yellow-green light, its spectral distribution is about that of a theoretical blackbody at 5900°K. For other ultraviolet and infrared wavelengths, the best fitting blackbody curves vary from 4500°K to 5300°K. (However, in the microwave region, where radiation comes from the solar corona, the best fit comes from a 2,000,000°K curve!)

Fig. 3-1, adapted from the handbook by Valley (19), shows the overall solar spectrum and illustrates the goodness of fit of the blackbody curve. It also shows, in a preliminary way, how this radiation is attenuated in the atmosphere. A more precise figure, showing solar radiation from 60 Å in the X-ray region to 6 metres in the radio region, may be found in Thekaekara (17). It should be noted that neither Fig. 3-1 nor the figure in Thekaekara illustrates the great complexity caused by the Fraunhofer spectrum, which consists of fine absorption lines superimposed on the continuum.

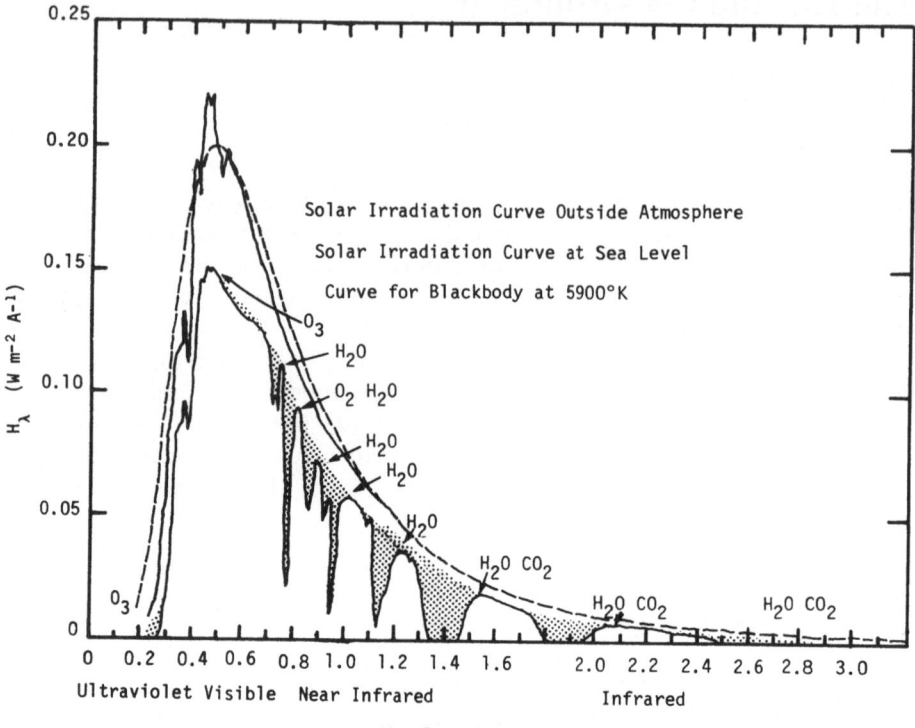

Fig. 3-1. Solar Irradiation. Modified from Valley to show the overall solar spectrum and illustrates the goodness of fit of the blackbody curve.

2. Effect of the Atmosphere

Our atmosphere attenuates sunlight by several mechanisms, which can be conveniently classified as Raleigh scattering, aerosol scattering and absorption by atomic and molecular constituents of the air. Gates (6) gives a practical introductory discussion of the three. Overall, since there is less air to pass through, we may say that more solar radiation reaches the surface at high altitudes than at low altitudes.

The Raleigh law of molecular scattering relates wave length of radiation, number of molecules in the air, scattering angle, and many other variables. Scattering is inversely dependent on the fourth power of the wavelength; blue and ultraviolet wavelengths are more scattered than red and infrared.

Aerosol scattering is not as easily treated. Its theoretical treatment is complex, but in the practical case, the greatest difficulty occurs in specifying the size-distribution model of aerosol particles in the atmosphere and their basic optical properties. Most theoretical treatments assume one or another vertical distribution of particles of aerosols in the atmosphere as a standard for calculations. Of course, the actual distribution will vary with time and place, and variations will certainly be greatest near the surface.

In this chapter, surface concentrations of aerosols (and hence the vertical aerosol profile) are assumed *not* to vary with the altitude of the earth's

36

surface. This implies that sea level models can be applied to the highlands, since aerosol concentrations are not sensitive to the different atmospheric conditions in the highlands. This is only a good first approximation, but it is not completely true. Still, use of upper air aerosol values would be much less realistic.

There are relatively few absorbing species in the near ultraviolet-visible-near infrared region. Ozone absorbs most of the sunlight from 2000 to 3000 Å, as well as a very small amount in its visible bands. Water vapour absorbs some near-infrared radiation, as do oxygen and carbon dioxide (3). Fig. 3-1 shows the practical extent of these absorption bands.

In addition, molecular oxygen absorbs strongly below 2000 Å, and various atomic species, particularly O and N, absorb below 1000 Å. Water absorbs strongly in various bands scattered through the infrared region. Many other molecules in the atmosphere, such as CO, CH_4 and H_2S, also absorb infrared radiation (3). However, our prime concern is the near ultraviolet-visible-near infrared 'window' from 3000 Å to about 15,000 Å (0.3 μm to 1.5 μm).

Superimposed on these factors are the effects of differences in the latitude of the laboratory, time of day and season, all of which affect the angle of elevation of the sun. Fig. 3-2 demonstrates the strong effect of this angle on the spectral composition of direct sunlight (13).

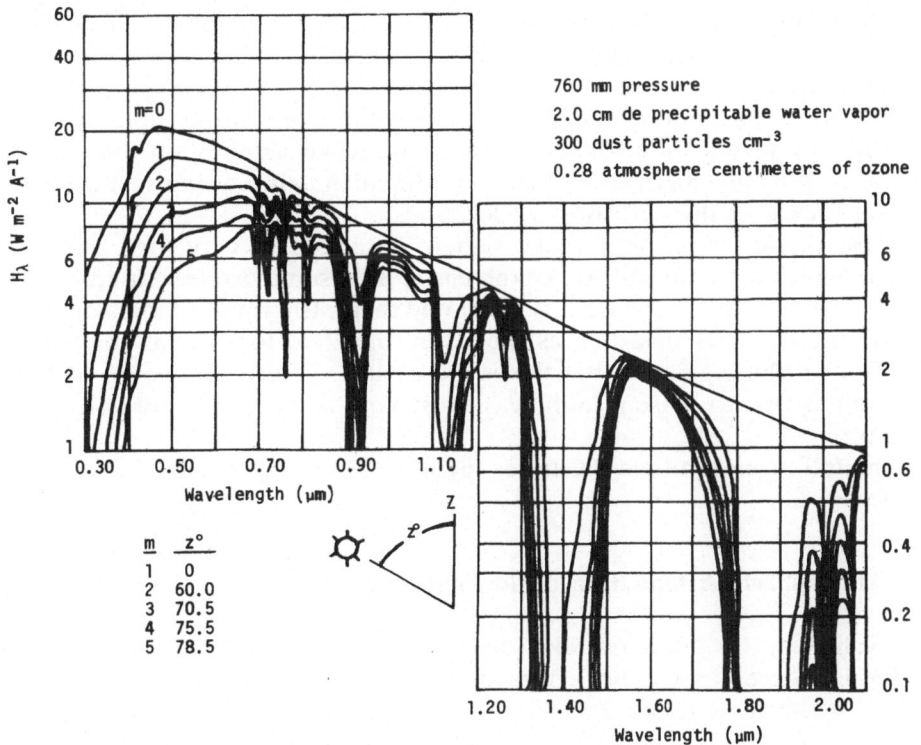

Fig. 3-2. Angular dependence of solar irradiation.

37

In practice, correction for sun elevation is normally made as the secant of the zenith angle (angle from the vertical, z). This is usually called the 'air mass', at least at sea level. (Under certain conditions an 'absolute air mass' is used for reduced pressure conditions; however, this is *not* used here.) For zenith angles over about 70°, air masses from Bemporad's tables (19) should be substituted for secants; these correct for refraction and the curvature of the earth, both of which present problems when the sun is very low in the sky.

3. Sunlight that Reaches Highlands Directly

The spectral composition of sunlight outside the atmosphere is presented by Thekaekara (14). His values, summarized in Table 3-1, represent the most recent consensus. They have been adopted by NASA as design criteria for space vehicles, and they are under review for approval by the joint ASTM/AIAA/IES* committee as an engineering standard.

Elterman has prepared extensive tables of attenuation of sunlight by a clear atmosphere (3). He listed the attenuation coefficients for each of the three attenuating mechanisms. Each table represents a certain region of the spectrum, and each lists the coefficients for various altitudes above sea level.

The tables presented here as Table 3-2 are a modification of Elterman's original tables. They use his sea level aerosol coefficients for all surface elevations. Since the external optical thickness is simply the sum of the Raleigh, aerosol and ozone coefficients, this is a simple modification. Elterman's upper air parameters are included in the tables for completeness.

The only justification for this modification is our assumption that surface turbulence is the chief cause of aerosol formation and that the surface is the actual source of the aerosol particles.

Most of the ozone in the atmosphere is limited to very high altitudes, much higher than any surface of the earth. Still, some ozone is found at the surface of the earth because it is generated there. Elterman's tables could be modified to reflect this, just as they were modified to account for surface aerosols in the highlands, but this is too small an effect to concern us here. Variations due to season, latitude, hemisphere, and other factors are much larger.

The following example illustrates the use of Elterman's and Thekaekara's tables.

4. Atmospheric Attenuation of Sunlight on a Clear Day

All values in this table are taken from Elterman's 1968 tables (4), with the following exceptions. The values of external optical thickness above sea level are calculated with underlined values of aerosol optical thickness, and

---

*American Society for Testing and Materials/American Institute of Aeronautics and Astronautics/Institute of Environmental Sciences.

Table 3-1. Solar radiation outside of the atmosphere.

λ	P_λ	D_λ	λ	P_λ	D_λ	λ	P_λ	D_λ
0.12	0.01	4×10^{-4}	0.39	109.8	7.82	0.75	123.5	51.69
0.14	0.003	5×10^{-4}	0.40	142.9	8.73	0.80	110.9	56.02
0.16	0.023	7×10^{-4}	0.41	175.1	9.92	0.90	89.1	63.37
0.18	0.125	1.7×10^{-3}	0.42	174.7	11.22	1.00	74.8	69.49
0.20	1.07	8.1×10^{-3}	0.43	163.9	12.47	1.2	48.5	78.40
0.22	5.75	0.05	0.44	181.0	13.73	1.4	33.7	84.33
0.23	6.67	0.10	0.45	200.6	15.14	1.6	24.5	88.61
0.24	6.30	0.14	0.46	206.6	16.65	1.8	15.9	91.59
0.25	7.04	0.19	0.47	203.3	18.17	2.0	10.3	93.49
0.26	13.0	0.27	0.48	207.4	19.68	2.2	7.9	94.83
0.27	23.2	0.41	0.49	195.0	21.15	2.4	6.2	95.86
0.275	20.4	0.48	0.50	194.2	22.60	2.6	4.8	96.67
0.28	22.2	0.56	0.51	188.2	24.01	2.8	3.9	97.31
0.285	31.5	0.66	0.52	183.3	25.38	3.0	3.1	97.83
0.29	48.2	0.81	0.53	184.2	26.74	3.2	2.26	98.22
0.295	58.4	1.01	0.54	178.3	28.08	3.4	1.66	98.50
0.30	51.4	1.21	0.55	172.5	29.38	3.6	1.35	98.72
0.305	60.3	1.42	0.56	169.5	30.65	3.8	1.11	98.91
0.31	68.9	1.66	0.57	171.2	31.91	4.0	0.95	99.06
0.315	76.4	1.92	0.58	171.5	33.18	4.5	0.59	99.34
0.32	83.0	2.22	0.59	170.0	34.44	5.0	0.38	99.51
0.325	97.5	2.55	0.60	166.6	35.68	6.0	0.18	99.72
0.33	105.9	2.93	0.62	160.2	38.10	7.0	0.10	99.82
0.34	107.4	3.72	0.64	154.4	40.42	8.0	0.06	99.88
0.35	109.3	4.52	0.66	148.6	42.66	10.0	0.025	99.94
0.36	106.8	5.32	0.68	142.7	44.81	15.0	4.9×10^{-3}	99.98
0.37	118.1	6.15	0.70	136.9	48.88	20.0	1.6×10^{-3}	99.99
0.38	112.0	7.0	0.72	131.4	48.86	50.0	3.8×10^{-5}	100.00

λ – Wavelength (μm).
P – Solar spectral irradiance averaged over small bandwidth centred at λ (mW cm^{-2} μm^{-1}).
D_λ – Percentage of the solar constant associated with wavelengths shorter than λ. Solar Constant – 135.30 mW cm^{-2}.

are therefore different from the original. The values in parentheses are useful for upper air but not for high surface elevations.

At a surface elevation of 4000 m, when the sun makes a 30° angle from the vertical, how much light arrives in the spectral interval from 3600 to 4000 Ångstrom units (0.36 μm to 0.40 μm)? How does this compare with sea level?

The total energy outside the atmosphere in this spectral region is:

$$\frac{8.73 - 5.32}{100} \times 135.3 = 4.61 \text{ mW/cm}^2$$

using data from Table 3-1 and the currently accepted value of the solar constant.

The external optical thickness, τ, is 0.638 ($= 0.274 + 0.364 + 0.000$) at 4000 metres, but $\tau = 0.814$ ($= 0.450 + 0.364 + 0.000$) at sea level, using

Table 3-2

	Parameters at 0.27 microns					Parameters at 0.28 microns			
Alti-tude kilo-metres h	Rayleigh optical thickness $(h-\infty)$ τ'_r	Aerosol optical thickness $(h-\infty)$ τ'_p	Ozone optical thickness $(h-\infty)$ τ'_3	External optical thickness $(h-\infty)$ $\supset'_{ext}$	Alti-tude kilo-metres h	Rayleigh optical thickness $(h-\infty)$ τ'_r	Aerosol optical thickness $(h-\infty)$ τ'_p	Ozone optical thickness $(h-\infty)$ τ'_3	External optical thickness $(h-\infty)$ τ'_{ext}
0	1.928	0.458	70.956	73.346	0	1.645	0.427	35.816	37.891
1	1.710	(0.250)	70.240	72.408(72.203)	1	1.460	(0.232)	35.455	37.342(37.149)
2	1.513	(0.158)	69.590	71.561(71.265)	2	1.291	(0.147)	35.126	36.844(36.568)
3	1.335	(0.119)	69.020	70.813(70.477)	3	1.139	(0.111)	34.839	36.405(36.091)
4	1.174	(0.102)	68.520	70.152(69.799)	4	1.002	(0.095)	34.586	36.015(35.685)
5	1.030	(0.091)	68.051	69.539(69.174)	5	0.879	(0.085)	34.350	35.656(35.315)
6	0.900	(0.083)	67.592	68.950(68.577)	6	0.768	(0.077)	34.118	35.313(34.965)
7	0.783	(0.077)	67.131	68.372(67.993)	7	0.669	(0.071)	33.885	34.981(34.627)
8	0.680	(0.071)	66.658	67.796(67.410)	8	0.580	(0.066)	33.645	34.653(34.293)
9	0.587	(0.065)	66.123	67.168(66.777)	9	0.501	(0.060)	33.376	34.304(33.939)

	Parameters at 0.34 microns					Parameters at 0.36 microns			
Alti-tude kilo-metres h	Rayleigh optical thickness $(h-\infty)$ τ'_r	Aerosol optical thickness $(h-\infty)$ τ'_p	Ozone optical thickness $(h-\infty)$ τ'_3	External optical thickness $(h-\infty)$ $\supset'_{ext}$	Alti-tude kilo-metres h	Rayleigh optical thickness $(h-\infty)$ τ'_r	Aerosol optical thickness $(h-\infty)$ τ'_p	Ozone optical thickness $(h-\infty)$ τ'_3	External optical thickness $(h-\infty)$ τ'_{ext}
0	0.717	0.379	0.022	1.120	0	0.564	0.379	0.001	0.945
1	0.636	(0.207)	0.021	1.036(0.866)	1	0.501	(0.207)	0.001	0.881(0.709)
2	0.563	(0.131)	0.021	0.963(0.717)	2	0.443	(0.131)	0.001	0.823(0.575)
3	0.497	(0.099)	0.021	0.897(0.618)	3	0.391	(0.099)	0.001	0.771(0.491)
4	0.437	(0.084)	0.021	0.837(0.543)	4	0.344	(0.084)	0.001	0.724(0.429)
5	0.383	(0.075)	0.021	0.783(0.480)	5	0.301	(0.075)	0.001	0.681(0.378)
6	0.335	(0.069)	0.021	0.735(0.425)	6	0.263	(0.069)	0.001	0.643(0.333)
7	0.292	(0.063)	0.020	0.691(0.376)	7	0.229	(0.063)	0.001	0.609(0.294)
8	0.253	(0.058)	0.020	0.692(0.332)	8	0.199	(0.058)	0.001	0.579(0.258)
9	0.219	(0.053)	0.020	0.618(0.293)	9	0.172	(0.053)	0.001	0.552(0.226)

	Parameters at 0.45 microns					Parameters at 0.50 microns			
Alti-tude kilo-metres h	Rayleigh optical thickness $(h-\infty)$ τ'_r	Aerosol optical thickness $(h-\infty)$ τ'_p	Ozone optical thickness $(h-\infty)$ τ'_3	External optical thickness $(h-\infty)$ $\supset'_{ext}$	Alti-tude kilo-metres h	Rayleigh optical thickness $(h-\infty)$ τ'_r	Aerosol optical thickness $(h-\infty)$ τ'_p	Ozone optical thickness $(h-\infty)$ τ'_3	External optical thickness $(h-\infty)$ τ'_{ext}
0	0.223	0.285	0.001	0.509	0	0.145	0.264	0.012	0.421
1	0.198	(0.155)	0.001	0.484(0.355)	1	0.129	(0.144)	0.012	0.405(0.284)
2	0.175	(0.098)	0.001	0.461(0.275)	2	0.114	(0.091)	0.011	0.389(0.217)
3	0.155	(0.074)	0.001	0.441(0.230)	3	0.100	(0.069)	0.011	0.375(0.181)
4	0.136	(0.063)	0.001	0.422(0.201)	4	0.088	(0.058)	0.011	0.363(0.158)
5	0.119	(0.056)	0.001	0.405(0.177)	5	0.077	(0.052)	0.011	0.352(0.141)
6	0.104	(0.052)	0.001	0.390(0.157)	6	0.068	(0.048)	0.011	0.343(0.127)
7	0.091	(0.048)	0.001	0.377(0.140)	7	0.059	(0.044)	0.011	0.334(0.114)
8	0.079	(0.044)	0.001	0.365(0.124)	8	0.051	(0.041)	0.011	0.326(0.103)
9	0.068	(0.040)	0.001	0.354(0.109)	9	0.044	(0.037)	0.011	0.319(0.092)

Table 3-2 (*Continued*)

Parameters at 0.30 microns

Altitude kilometres h	Rayleigh optical thickness $(h-\infty)$ τ'_r	Aerosol optical thickness $(h-\infty)$ τ'_p	Ozone optical thickness $(h-\infty)$ τ'_3	External optical thickness $(h-\infty)$ τ'_{ext}
0	1.222	0.411	3.413	5.047
1	1.084	(0.224)	3.373	4.873(4.688)
2	0.959	(0.142)	3.347	4.717(4.450)
3	0.846	(0.107)	3.320	4.577(4.274)
4	0.744	(0.091)	3.295	4.450(4.132)
5	0.652	(0.081)	3.273	4.336(4.008)
6	0.570	(0.074)	3.251	4.232(3.897)
7	0.496	(0.069)	3.229	4.136(3.795)
8	0.431	(0.063)	3.206	(3.701)
9	0.372	(0.058)	3.180	(3.611)

Parameters at 0.32 microns

Altitude kilometres h	Rayleigh optical thickness $(h-\infty)$ τ'_r	Aerosol optical thickness $(h-\infty)$ τ'_p	Ozone optical thickness $(h-\infty)$ τ'_3	External optical thickness $(h-\infty)$ τ'_{ext}
0	0.927	0.395	0.303	1.628
1	0.823	(0.215)	0.300	1.518(1.340)
2	0.728	(0.136)	0.298	1.421(1.163)
3	0.642	(0.103)	0.295	1.332(1.041)
4	0.565	(0.088)	0.293	1.253(0.947)
5	0.495	(0.078)	0.291	1.181(0.866)
6	0.433	(0.072)	0.289	1.117(0.794)
7	0.377	(0.066)	0.287	1.059(0.731)
8	0.327	(0.061)	0.285	1.007(0.674)
9	0.283	(0.056)	0.283	0.961(0.622)

Parameters at 0.38 microns

Altitude kilometres h	Rayleigh optical thickness $(h-\infty)$ τ'_r	Aerosol optical thickness $(h-\infty)$ τ'_p	Ozone optical thickness $(h-\infty)$ τ'_3	External optical thickness $(h-\infty)$ τ'_{ext}
0	0.450	0.364	0.000	0.814
1	0.399	(0.198)	0.000	0.763(0.598)
2	0.353	(0.126)	0.000	0.717(0.480)
3	0.312	(0.095)	0.000	0.676(0.407)
4	0.274	(0.081)	0.000	0.638(0.355)
5	0.240	(0.072)	0.000	0.604(0.313)
6	0.210	(0.066)	0.000	0.574(0.276)
7	0.183	(0.061)	0.000	0.547(0.244)
8	0.152	(0.056)	0.000	0.523(0.215)
9	0.137	(0.051)	0.000	0.501(0.189)

Parameters at 0.40 microns

Altitude kilometres h	Rayleigh optical thickness $(h-\infty)$ τ'_r	Aerosol optical thickness $(h-\infty)$ τ'_p	Ozone optical thickness $(h-\infty)$ τ'_3	External optical thickness $(h-\infty)$ τ'_{ext}
0	0.364	0.316	0.000	0.680
1	0.323	(0.172)	0.000	0.639(0.495)
2	0.285	(0.109)	0.000	0.601(0.395)
3	0.252	(0.082)	0.000	0.568(0.335)
4	0.221	(0.070)	0.000	0.537(0.292)
5	0.194	(0.063)	0.000	0.510(0.257)
6	0.170	(0.057)	0.000	0.486(0.227)
7	0.148	(0.053)	0.000	0.464(0.201)
8	0.128	(0.049)	0.000	0.444(0.177)
9	0.111	(0.044)	0.000	0.427(0.156)

Parameters at 0.55 microns

Altitude kilometres h	Rayleigh optical thickness $(h-\infty)$ τ'_r	Aerosol optical thickness $(h-\infty)$ τ'_p	Ozone optical thickness $(h-\infty)$ τ'_3	External optical thickness $(h-\infty)$ τ'_{ext}
0	0.098	0.250	0.031	0.379
1	0.087	(0.136)	0.031	0.368(0.254)
2	0.077	(0.086)	0.030	0.357(0.194)
3	0.068	(0.065)	0.030	0.348(0.163)
4	0.060	(0.055)	0.030	0.340(0.145)
5	0.052	(0.049)	0.030	0.332(0.132)
6	0.046	(0.045)	0.030	0.326(0.121)
7	0.040	(0.042)	0.029	0.319(0.111)
8	0.35	(0.038)	0.029	0.314(0.102)
9	0.030	(0.035)	0.029	0.309(0.094)

Parameters at 0.60 microns

Altitude kilometres h	Rayleigh optical thickness $(h-\infty)$ τ'_r	Aerosol optical thickness $(h-\infty)$ τ'_p	Ozone optical thickness $(h-\infty)$ τ'_3	External optical thickness $(h-\infty)$ τ'_{ext}
0	0.069	0.237	0.045	0.351
1	0.061	(0.129)	0.044	0.342(0.235)
2	0.054	(0.082)	0.044	0.335(0.180)
3	0.048	(0.062)	0.043	0.328(0.153)
4	0.042	(0.053)	0.043	0.322(0.138)
5	0.037	(0.047)	0.043	0.317(0.127)
6	0.032	(0.043)	0.042	0.311(0.118)
7	0.028	(0.040)	0.042	0.307(0.110)
8	0.024	(0.037)	0.042	0.303(0.103)
9	0.021	(0.033)	0.042	0.300(0.096)

Table 3-2 (*Continued*)

Parameters at 0.65 microns | Parameters at 0.70 microns

Altitude kilometres h	Rayleigh optical thickness $(h-\infty)$ τ'_r	Aerosol optical thickness $(h-\infty)$ τ'_p	Ozone optical thickness $(h-\infty)$ τ'_3	External optical thickness $(h-\infty)$ $\supset'_{ext}$	Altitude kilometres h	Rayleigh optical thickness $(h-\infty)$ τ'_r	Aerosol optical thickness $(h-\infty)$ τ'_p	Ozone optical thickness $(h-\infty)$ τ'_3	External optical thickness $(h-\infty)$ τ'_{ext}
0	0.050	0.224	0.021	0.295	0	0.037	0.213	0.008	0.258
1	0.044	(0.122)	0.021	0.289(0.187)	1	0.033	(0.116)	0.008	0.254(0.157)
2	0.039	(0.078)	0.021	0.284(0.137)	2	0.029	(0.074)	0.008	0.250(0.110)
3	0.034	(0.058)	0.020	0.278(0.113)	3	0.026	(0.056)	0.008	0.247(0.089)
4	0.030	(0.050)	0.020	0.274(0.100)	4	0.022	(0.047)	0.008	0.243(0.077)
5	0.027	(0.044)	0.020	0.271(0.091)	5	0.020	(0.042)	0.007	0.240(0.069)
6	0.023	(0.041)	0.020	0.267(0.084)	6	0.017	(0.039)	0.007	0.237(0.063)
7	0.020	(0.038)	0.020	0.264(0.078)	7	0.015	(0.036)	0.007	0.235(0.058)
8	0.018	(0.035)	0.020	0.262(0.072)	8	0.013	(0.033)	0.007	0.233(0.053)
9	0.015	(0.032)	0.020	0.259(0.066)	9	0.011	(0.030)	0.007	0.231(0.049)

Parameters at 1.06 microns | Parameters at 1.26 microns

Altitude kilometres h	Rayleigh optical thickness $(h-\infty)$ τ'_r	Aerosol optical thickness $(h-\infty)$ τ'_p	Ozone optical thickness $(h-\infty)$ τ'_3	External optical thickness $(h-\infty)$ $\supset'_{ext}$	Altitude kilometres h	Rayleigh optical thickness $(h-\infty)$ τ'_r	Aerosol optical thickness $(h-\infty)$ τ'_p	Ozone optical thickness $(h-\infty)$ τ'_3	External optical thickness $(h-\infty)$ τ'_{ext}
0	0.007	0.179	0.000	0.186	0	0.003	0.171	0.000	0.174
1	0.006	(0.097)	0.000	0.185(0.103)	1	0.003	(0.093)	0.000	0.174(0.096)
2	0.005	(0.062)	0.000	0.184(0.067)	2	0.003	(0.059)	0.000	0.174(0.062)
3	0.005	(0.046)	0.000	0.184(0.051)	3	0.002	(0.044)	0.000	0.173(0.047)
4	0.004	(0.040)	0.000	0.183(0.044)	4	0.002	(0.038)	0.000	0.173(0.040)
5	0.004	(0.035)	0.000	0.183(0.039)	5	0.002	(0.034)	0.000	0.173(0.036)
6	0.003	(0.032)	0.000	0.182(0.036)	6	0.002	(0.031)	0.000	0.173(0.033)
7	0.003	(0.030)	0.000	0.182(0.033)	7	0.001	(0.029)	0.000	0.172(0.030)
8	0.002	(0.028)	0.000	0.181(0.030)	8	0.001	(0.026)	0.000	0.172(0.028)
9	0.002	(0.025)	0.000	0.181(0.027)	9	0.001	(0.024)	0.000	0.172(0.025)

Parameters at 3.50 microns | Parameters at 4.00 microns

Altitude kilometres h	Rayleigh optical thickness $(h-\infty)$ τ'_r	Aerosol optical thickness $(h-\infty)$ τ'_p	Ozone optical thickness $(h-\infty)$ τ'_3	External optical thickness $(h-\infty)$ $\supset'_{ext}$	Altitude kilometres h	Rayleigh optical thickness $(h-\infty)$ τ'_r	Aerosol optical thickness $(h-\infty)$ τ'_p	Ozone optical thickness $(h-\infty)$ τ'_3	External optical thickness $(h-\infty)$ τ'_{ext}
0	0.000	0.111	0.000	0.111	0	0.000	0.100	0.000	0.100
1	0.000	(0.060)	0.000	0.111(0.060)	1	0.000	(0.054)	0.000	0.100(0.054)
2	0.000	(0.038)	0.000	0.111(0.038)	2	0.000	(0.034)	0.000	0.100(0.034)
3	0.000	(0.029)	0.000	0.111(0.029)	3	0.000	(0.026)	0.000	0.100(0.026)
4	0.000	(0.025)	0.000	0.111(0.025)	4	0.000	(0.022)	0.000	0.100(0.022)
5	0.000	(0.022)	0.000	0.111(0.022)	5	0.000	(0.020)	0.000	0.100(0.020)
6	0.000	(0.020)	0.000	0.111(0.020)	6	0.000	(0.018)	0.000	0.100(0.018)
7	0.000	(0.019)	0.000	0.111(0.019)	7	0.000	(0.017)	0.000	0.100(0.017)
8	0.000	(0.017)	0.000	0.111(0.017)	8	0.000	(0.015)	0.000	0.100(0.15)
9	0.000	(0.016)	0.000	0.111(0.016)	9	0.000	(0.014)	0.000	0.100(0.014)

Table 3-2 (*Continued*)

	Parameters at 0.80 microns					Parameters at 0.90 microns			
Alti-tude kilo-metres h	Rayleigh optical thickness $(h-\infty)$ τ_r'	Aerosol optical thickness $(h-\infty)$ τ_p'	Ozone optical thickness $(h-\infty)$ τ_3'	External optical thickness $(h-\infty)$ $\supset_{ext}'$	Alti-tude kilo-metres h	Rayleigh optical thickness $(h-\infty)$ τ_r'	Aerosol optical thickness $(h-\infty)$ τ_p'	Ozone optical thickness $(h-\infty)$ τ_3'	External optical thickness $(h-\infty)$ τ_{ext}'
0	0.021	0.201	0.003	0.226	0	0.013	0.190	0.000	0.203
1	0.019	(0.109)	0.003	0.223(0.132)	1	0.012	(0.103)	0.000	0.202(0.115)
2	0.017	(0.069)	0.003	0.221(0.090)	2	0.010	(0.066)	0.000	0.200(0.076)
3	0.015	(0.052)	0.003	0.219(0.070)	3	0.009	(0.049)	0.000	0.199(0.059)
4	0.013	(0.044)	0.003	0.217(0.061)	4	0.008	(0.042)	0.000	0.198(0.050)
5	0.011	(0.040)	0.003	0.215(0.055)	5	0.007	(0.038)	0.000	0.197(0.045)
6	0.010	(0.036)	0.003	0.214(0.050)	6	0.006	(0.034)	0.000	0.196(0.041)
7	0.009	(0.034)	0.003	0.213(0.046)	7	0.005	(0.032)	0.000	0.195(0.037)
8	0.008	(0.031)	0.003	0.212(0.042)	8	0.005	(0.029)	0.000	0.195(0.034)
9	0.007	(0.028)	0.003	0.211(0.036)	9	0.004	(0.027)	0.000	0.194(0.031)

	Parameters at 1.67 microns					Parameters at 2.17 microns			
Alti-tude kilo-metres h	Rayleigh optical thickness $(h-\infty)$ τ_r'	Aerosol optical thickness $(h-\infty)$ τ_p'	Ozone optical thickness $(h-\infty)$ τ_3'	External optical thickness $(h-\infty)$ $\supset_{ext}'$	Alti-tude kilo-metres h	Rayleigh optical thickness $(h-\infty)$ τ_r'	Aerosol optical thickness $(h-\infty)$ τ_p'	Ozone optical thickness $(h-\infty)$ τ_3'	External optical thickness $(h-\infty)$ τ_{ext}'
0	0.001	0.155	0.000	0.156	0	0.000	0.134	0.000	0.135
1	0.001	(0.084)	0.000	0.156(0.085)	1	0.000	(0.073)	0.000	0.134(0.074)
2	0.001	(0.054)	0.000	0.156(0.054)	2	0.000	(0.046)	0.000	0.134(0.047)
3	0.001	(0.040)	0.000	0.156(0.041)	3	0.000	(0.035)	0.000	0.134(0.035)
4	0.001	(0.034)	0.000	0.156(0.035)	4	0.000	(0.030)	0.000	0.134(0.030)
5	0.001	(0.031)	0.000	0.156(0.031)	5	0.000	(0.027)	0.000	0.134(0.027)
6	0.001	(0.028)	0.000	0.156(0.029)	6	0.000	(0.024)	0.000	0.134(0.025)
7	0.000	(0.026)	0.000	0.155(0.026)	7	0.000	(0.022)	0.000	0.134(0.023)
8	0.000	(0.024)	0.000	0.155(0.024)	8	0.000	(0.021)	0.000	0.134(0.021)
9	0.000	(0.022)	0.000	0.155(0.022)	9	0.000	(0.019)	0.000	0.134(0.019)

data for 0.38 μm from Table 3-2. Therefore, the transmission coefficient, $T = \exp(-\tau \sec z) = \exp(-0.638 \sec 30°) = 0.480$ at 4000 metres but $T = \exp(-0.814 \sec 30°) = 0.391$ at sea level. Therefore, $0.458 \times 4.61 = 2.1 \text{ mW/cm}^2$ of this light arrives at 4000 m but only $0.391 \times 4.61 = 1.8 \text{ mW/cm}^2$ arrive at sea level.

This calculation illustrates the simple use of these tables, but we should note that the tables reproduced here are very drastic abridgements of the originals. The original of Table 3-1 is tabulated for much narrower spectral intervals, as well as for many other functions of radiance. Table 3-2 in the original allows calculations to be made between two altitudes and between two horizontal points at any given altitude.

For completeness, Elterman's tables extend from 2700 Å in the ultraviolet to 40,000 Å (4 μm) in the infrared. However, light has not been measured anywhere on the surface of the earth at wavelengths shorter than 2863 Å (9), and there are many absorption bands in the near infrared and infrared. Use of these data below 2900 Å poses no problem, but much care is required above 6000 Å. These tables are only concerned with atmospheric windows in this region, and therefore, interpolation to intermediate wavelengths, which would lie within atmospheric absorption bands instead of between them, would be very deceptive.

5. The Effect of Haze

Sunlight does not penetrate haze as well as in clear atmospheres. Since haze is often due to water particles of a certain size distribution, its attenuating power can be calculated in an approximate way.

Elterman also prepared a family of models for hazy atmospheres (4). He used the 'meteorological range', roughly equal to visual range, as his measure of haziness, and he assumed an aerosol mixing layer of five km. His Raleigh and ozone parameters are the same in these models as for his clear atmosphere model, mentioned before.

Elterman concludes that a range of about 13 km separates clear atmospheres from hazy ones. That is, if one can distinguish mountains farther away than 13 km, he may consider the day to be clear. He also holds that a range of 1.2 km separates haze from fog; it is a foggy day if a person can only distinguish objects closer than a kilometre. Between these two limits lie various degrees of haziness.

Again, the original model only contemplates land at sea level. To adapt these tables for the highlands we must again assume that his surface conditions will be valid for surfaces in the highlands.

Elterman's surface parameters are presented in Table 3-3, as functions both of meteorological range and wavelength. Their use is clarified by the following example.

Conditions are the same as in the previous example, except that because of haze a person cannot distinguish features more distant than five km. Answer the same questions as before.
From Table 3-3, for 0.38 μm with a range of 5 km, τ for aerosols is 1.174. Therefore, replace 0.638 with 1.448 ($= 0.274 + 1.174 + 0.000$) for 4000 metres and replace 0.814 with 1.624 ($= 0.450 + 1.174 + 0.000$) for sea level. Proceeding as before yields 0.87 mW/cm² for 4000 m and 0.70 mW/cm² for sea level.

6. Total Direct Radiation

Tables 3-2 and 3-3 are not very useful for calculating the total intensity of the direct radiation reaching the earth. Both deal with window portions of the spectrum, and they do not extend far into the infrared region.

Table 3-3. Attenuation of sunlight as a function of meteorological range.

SPECTRAL REGION, τ'_p(Microns)

Meteoro-logical range, km	0.27	0.28	0.30	0.32	0.34	0.36	0.38	0.40	0.45	0.50
2	3.392	3.197	3.023	2.849	2.675	2.521	2.431	2.229	1.978	1.812
3	2.440	2.299	2.175	2.050	1.926	1.818	1.752	1.605	1.425	1.305
4	1.941	1.828	1.730	1.632	1.534	1.449	1.397	1.278	1.135	1.040
5	1.629	1.534	1.452	1.370	1.288	1.219	1.174	1.073	0.954	0.874
6	1.415	1.332	1.262	1.191	1.120	1.061	1.022	0.933	0.829	0.760
8	1.136	1.069	1.013	0.956	0.900	0.854	0.823	0.750	0.667	0.611
10	0.963	0.906	0.859	0.812	0.764	0.726	0.699	0.637	0.566	0.520
13	0.797	0.749	0.711	0.672	0.633	0.602	0.580	0.528	0.469	0.431
	0.55	0.60	0.65	0.70	0.80	0.90	1.06	1.26	1.67	2.17
2	1.676	1.510	1.381	1.297	1.153	1.053	0.954	0.872	0.752	0.672
3	1.208	1.090	0.998	0.937	0.834	0.763	0.692	0.633	0.547	0.488
4	0.963	0.870	0.797	0.749	0.667	0.611	0.554	0.508	0.439	0.392
5	0.809	0.732	0.671	0.630	0.563	0.515	0.468	0.429	0.372	0.331
6	0.704	0.638	0.585	0.550	0.491	0.450	0.409	0.375	0.325	0.290
8	0.567	0.514	0.472	0.444	0.397	0.364	0.331	0.305	0.265	0.235
10	0.482	0.437	0.402	0.378	0.339	0.311	0.283	0.261	0.227	0.202
13	0.400	0.364	0.334	0.315	0.283	0.260	0.237	0.219	0.190	0.169

To see the variation of total direct solar radiation, study the first and second columns of Table 3-4. These data were extracted from some monumental studies in the Eastern Alps at about 48°N, reported by Dirmhirn (2) and Sauberer (16). These are average measurements, not calculations. Again, original data are more comprehensive, given for 200 m intervals of altitude, individually for all months of the year, and in tenths of cloudiness from clear to completely overcast days.

Although the methods of measuring such data are quite well understood, their approximate calculation is quite another matter. Several authors have worked on the problem (15), but Majumdar et al. (10) have presented a method that seems to be adequate for most uses of the non-specialist; however, his method yields only instantaneous values, not daily totals.

Majumdar's method is valid for a 'clear sky with minimal smoke and dust'. He assumes that the precipitable water for latitudes of India is about 0.16 p_w, following Hann's approximation. In a comparable way, one might use a world-wide average value of 2.0 cm, leading to a value of 0.25 p_w. Alternatively, p_w might be obtained from measurements of relative humidity combined with a table of the vapour pressure of water, or atmospheric water can be estimated from the data in Chapter II. In any case, calculated values of total radiation with Majumdar's method do not vary strongly with the assumption of any such reasonable value. More details on this method will not be given here.

As stated before, the above method yields only the instantaneous values of the total solar intensity. The problem of the total solar radiation reaching a given area is even more complex. The problem is reviewed by Koller (9), and Robinson (15) presents some useful tables (15).

7. Indirect Radiation

Diffuse sunlight comes mostly from Raleigh scattering, scattering by aerosols, sunlight reflected from clouds, and sunlight reflected from surface features above the horizon. It is strongly dependent on the angle of viewing and many other factors, and it cannot be summarized in any simple way. As a matter of fact, only in recent years have calculations been completed for scattering in the air, and these have been limited to certain assumed 'reasonable' models. Here again, Robinson presents some useful and pertinent tables (15).

Again, Table 3-4 illustrates some trends. On a completely overcast day, only sky radiation (also called diffuse radiation) can illuminate the surface of the earth. As expected, total indirect radiation increases with altitude on completely overcast days. However, indirect radiation decreases with altitude on clear days since there is less air present to cause Rayleigh scattering.

The spectral distribution of indirect radiation is yet another problem, so complex that we must content ourselves with only broad generalizations. Since Rayleigh scattering is very dependent on wavelength, indirect radiation is especially strong in ultraviolet. Indeed, Robinson states that below 3150 Ångstroms, it is more intense on a clear day than on a cloudless day (15). However, this is certainly not true at the longer visible and infrared wavelengths, which are scattered less.

As would be expected, direct radiation is greatly reduced by haze while diffuse radiation becomes stronger; this is shown quite well by the data in Table 3-4. Other important factors include surface topography, reflectivity of the surface, cloud cover, types of clouds and atmospheric composition, in addition to altitude. Dirmhirn (2) discusses these points at length; she shows very graphically how the lay of the land affects indirect radiation. For example, if the area surrounding the measuring station is flat and covered with snow, the downwelling scattered radiation will be much greater than if the station is located on an isolated peak.

Table 3-4. Direct and indirect solar radiation in the highlands. The numbers in this table are averages, with units of cal cm^{-2} day^{-1}.

Altitude in metres	Direct		Clear days Indirect		Total		Overcast days Indirect	
	June	December	June	December	June	December	June	December
200	592	101	99	29	691	130	155	30
1000	662	125	85	25	747	150	205	38
2000	728	144	71	22	799	166	293	54
3000	773	152	61	19	834	171	403	75

8. Total Radiation

In general, the global radiation reaching a given point is the sum of the direct and the sky radiation. All factors mentioned heretofore will affect global radiation. However, some effects will be unexpected.

For example, the total radiation reaching the earth is a complex function of cloud type and cloud disposition. Certain kinds of clouds can reduce light very severely. Other kinds can even increase the radiation reaching the surface, such that the contrast between sunny and shady areas is especially dramatic. Sunlight shining through the gaps may well be augmented by light reflected from the sides of clouds.

Turner (18) at 1940 m in the Eastern Alps, observed instantaneous radiation of $2.25\,cal\,cm^{-2}\,min^{-1}$, considerably greater than the solar constant. In addition, he observed sevenfold variations in intensity in one minute, eleven-fold variations in nine minutes, and fifteen-fold variations in eleven minutes. He also observed changes of the ground surface temperatures of ten degrees in just a few seconds, and found that surface temperatures in the upper reaches of the Alps were often higher than the hottest surface temperatures in the adjacent lowlands.

9. Limitations of Calculations

Despite the precision of Elterman's tables or, indeed, any other model of the optical properties of the atmosphere, it is currently not possible to make accurate calculations of solar radiation reaching the surface of the earth. Problems of such calculations are so severe that if a person needs accurate values of the solar radiation, he must measure them.

Our modification of Elterman's model presumes that the entire aerosol mixing layer is transposed upwards. That is, if the earth's surface is at 3000 m, then the top of the aerosol mixing layer is at 8000 m. This is probably not true, but the truth of the matter is not known. The net result is that calculations would probably yield values of transmission that are low, and therefore the values of the solar radiation may well be underestimated.

Elterman assumes a single vertical distribution of ozone, and this, again, is only a first approximation. Concentrations of ozone are always higher near the surface of the earth than, say, a kilometre above it. We have not modified these tables for this factor since it is minor compared with other sources of variation of ozone content.

For example, ozone is known to vary with the seasons, being normally highest in early spring and lowest in early fall in the northern hemisphere. It also varies with latitude, being normally lowest nearer the equator and greatest near the poles, and it is not identically distributed in the northern and southern hemispheres.

Ozone can also be formed through decomposition of nitrogen dioxide introduced into the atmosphere by factory smoke, forest fires, volcanic eruptions, vehicular traffic, etc. (The ozone content of smog in Los Angeles,

California, U.S.A., is said to have reached 0.5 ppm.) However, these variations of ozone will probably have little practical importance for the radiation environment except at the short wavelength limit of transmission of air.

Finally, we must recognize that all calculations are based on models, and only fortuitously could any model represent actual circumstances at any given time and place. Virtually all parameters of any model can vary in the real case.

10. Conclusion

Using methods and data from this chapter, a person can begin to understand the spectral distribution and the intensity of sunlight in the highlands. Exploratory calculations can be made and reasonable values may be obtained for the intensity of solar radiation under a few given conditions. However, if precise values of solar radiation are needed he must measure them.

11. References

1. Aldrich, L. B. & W. H. Hoover. 1954. The Solar Constant. *Ann. Astrophys. Lab.*, 7 (2): Smithsonian Inst. Washington DC.
2. Dirmhirn, I. 1951. Untersuchungen in der Himmelstrahlung in der Ostalpen mit besonderer Berücksichtigung ihrer Höhenabhängigkeit. *Arch. Meteorol. Geophys. Bioklimatol..* (G) 2 (4): 301–346.
3. Elterman, L. 1968. UV, visible and IR Attenuation for Altitudes to 50 km. AFCRL-68-0153, Env. Res. Paper, 285 Bedford. Mass.
4. Elterman, L. 1970. Vertical attenuation Model with eight surface meteorological ranges 2 to 13 kilometres. AFCRL-70-0200, Env. Res. Paper 318, Bedford. Mass.
5. Gates, D. M. 1960. *J. opt. Soc. Amer.*, 50: 1299 (See also in this connection E. J. McCartney, 1976. Optics of the Atmosphere, New York, John Wiley).
6. Gates, D. M. 1966. *Science*, 151: 523.
7. Gates, D. M. & W. J. Harrop, 1963. *Appl. Optics*, 2: 887.
8. Geiger, R. 1965. The Climate near the ground. Cambridge: Harvard Univ. Press.
9. Kollar, L. R. 1965. Ultraviolet radiation. New York: J. Wiley (Refers to the work F. W. P. Gotz, 1931. Strahlentner, 40: 690).
10. Majumdar, N. C., B. L. Mathur & S. B. Kaushik, 1971. Prediction of direct solar radiation for low atmospheric turbidity. *Solar Energy*, 13: 383–394.
11. McCartney, E. J. 1976. Optics of the Atmosphere. New York: John Wiley & Sons.
12. Middleton, W. E. Knowles, 1952. Vision through the atmosphere. University of Toronto.
13. Moon, P. 1940. *J. Franklin Inst.*, 230 (5): 583.
14. NASA Space Vehicle Design Criteria (Environment): Solar Electromagnetic Radiation. NASA SP-8005, rev. 1931, Washington.
15. Nathan, R. 1966. Solar Radiation. Elsevier, New York.
16. Sauberer, F. & I. Dirmhirn, 1960. Das Strahlungsklima. In: Klimatographie von Österreich, New York: Elsevier.
17. Thekaekara, M. P. 1973. *Solar Energy*, 14: 109.
18. Turner, H. 1958. Über das Licht- und Strahlungsklima einer Hanglage der Ötztaler Alpen bei Obergurgl und seine Auswirkung auf das Mikorklima und auf Vegetation. *Arch. Meteorol.*, (B) 8: 273–327.
19. Valley, S. 1963. Handbook of Geophysics and Space Environment. New York: McGraw-Hill Book Co.

Correcting boiling points

L. E. Giddings

If the lowering of atmospheric pressure is the most conspicuous of changes in the highlands environment, surely the lowering of boiling points is its most obvious effect. Eggs can't be hard-boiled as easily as at sea level, and handbook values of boiling points can't be used without correction.

Fortunately, because of its great practical importance, many distinguished scientists have devoted considerable efforts to correlating vapour pressure, and hence boiling point, with temperature. Dalton (1) explored the problem in 1801 and produced a formulation that had some value. One of the most useful of relations was published by Dühring (2) in 1878 and valuable articles still appear. We have an arsenal of methods at hand; and since we are interested here in a limited range of pressures, with only moderate precision, most methods give quite useful results.

1. Clausius–Clapeyron Equation

Virtually all methods with theoretical background ultimately depend on the Clapeyron equation:

$$\frac{dP}{dT} = \frac{L}{(V-v)T}. \tag{4.1}$$

In this equation, P represents pressure, T represents absolute temperature, L, the latent heat per mole, V, the volume of a mole of vapour, and v the volume of a mole of liquid. Since v is typically less than a thousandth of V it can conveniently be considered zero. This equation can be combined with the perfect gas law,

$$V = \frac{RT}{P}, \tag{4.2}$$

yielding the Clausius–Clapeyron equation:

$$\frac{dp}{dT} = \frac{LP}{RT^2} \quad \text{or} \quad \frac{dP}{P} = \frac{L}{RT^2} \, dT. \tag{4.3}$$

This can also be conveniently written in logarithmic form:

$$\frac{d \ln P}{dT} = \frac{L}{RT^2} \quad \text{or} \quad d \ln P = \frac{L}{RT^2} \, dT. \tag{4.4}$$

Its integrated form,

$$\ln P = -\frac{L}{RT} + C \quad \text{or} \quad \log P = -\frac{k}{T} + C, \tag{4.5}$$

49

is also used for vapour pressures, usually in the following form:

$$\ln \frac{P_2}{P_1} = -\frac{L}{R}\left(\frac{1}{T_2} - \frac{1}{T_1}\right) \quad \text{or} \quad \log \frac{P_2}{P_1} = -k\left(\frac{1}{T_2} - \frac{1}{T_1}\right). \qquad (4.6)$$

Although engineers and chemists tend to use the Clausius–Clapeyron equation in graphic form, as shown in the first example below, the second example shows that its direct use is not especially troublesome, except that slide rule accuracy is insufficient for most uses.

Example 4-1: The boiling point of diethyl ether is 34.6° and its vapour pressure is 100 mm at −11.5°. What would be its boiling point at 500 mm pressure? Solve this problem graphically.

First, convert all temperature data to absolute temperatures, then to their inverse values:

$P_1 = 760$ mm, $\quad t_1 = 34.6°, \quad T_1 = 307.8°K, \quad 1/T_1 = 0.003249285$

$P_2 = 100$ mm, $\quad t_2 = -11.5°, \quad T_2 = 261.7°K, \quad 1/T_2 = 0.003821753$

$P_3 = 500$ mm, $\quad t_3$ is to be determined.

With ordinary semilog paper, devise an acceptable division for $1/T$ on the linear scale, such as the one shown here (Fig. 4-1):

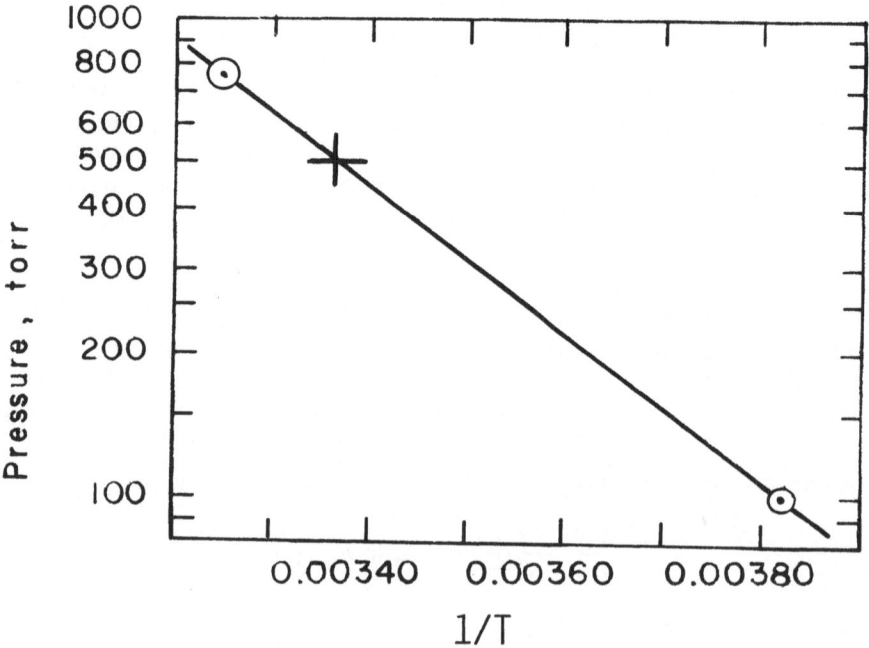

Fig. 4-1. Graphical solutions of boiling point of diethyl ether.

50

List pressure on the log scale, as shown here. Plot the two points and connect them with a straight line. From this graph, for 500 mm, $1/T_3$ will have a value of 0.003365. Therefore, $T_3 = 297.16$ and $t_3 = 24.0°$.

For this use, an ordinary sheet of one-cycle semilog paper will be very convenient. It will normally give sufficient precision for this sort of work.

It is obvious from this example that a scale reading directly in T or t could be constructed quite easily. Indeed, one commercial firm has prepared a series of such graph papers, a sample of which is included as Figure 4-2. However, since the intended use of their papers is for large ranges of pressures, they are not as useful as they might be. One must be prepared to make one's own graph.

Example 4-2: Solve the problem in example 4-1 using the Clausius–Clapeyron equation directly.

$$\log \frac{P_1}{P_2} = \frac{L}{2.303R} \left(\frac{1}{T_2} - \frac{1}{T_1} \right)$$

$$\log \frac{760}{100} = k(0.00382175 - 0.003249285)$$

where $k = L/2.303R = 1538.601$

$$\log \frac{P_1}{P_3} = 1538.601 \left(\frac{1}{T_3} - \frac{1}{T_1} \right)$$

$$\log \frac{760}{500} = 0.1818 = 1538.601 \left(\frac{1}{T_3} - 0.0032492851 \right)$$

$$T_3 = \frac{1}{0.003367444} = 296.96 \, °K$$

$$t_3 = 296.96 - 273.16 = 23.8°$$

2. Germann Charts

For use with pure organic liquids, as in the organic chemistry laboratory, Germann's (4) charts are probably the most convenient of all the practical formulations of the Clausius–Clapeyron equation. Germann prepared nomographs based on equation (4.6), using published values of vapour pressure as a function of temperature.

To the extent that liquids follow the Clausius–Clapeyron equation in the range of pressures found on the surface of the earth, Germann's charts are followed quite well. Their precision is well within a degree for most practical conditions.

The chief limitation of Germann's original charts is their lower limit of pressures, 500 torr. Several university and research laboratories are found at elevations above 4000 m, where the pressure would normally be about 450 torr. His charts, Fig. 4-3, and Table 4-1 need to be slightly modified to

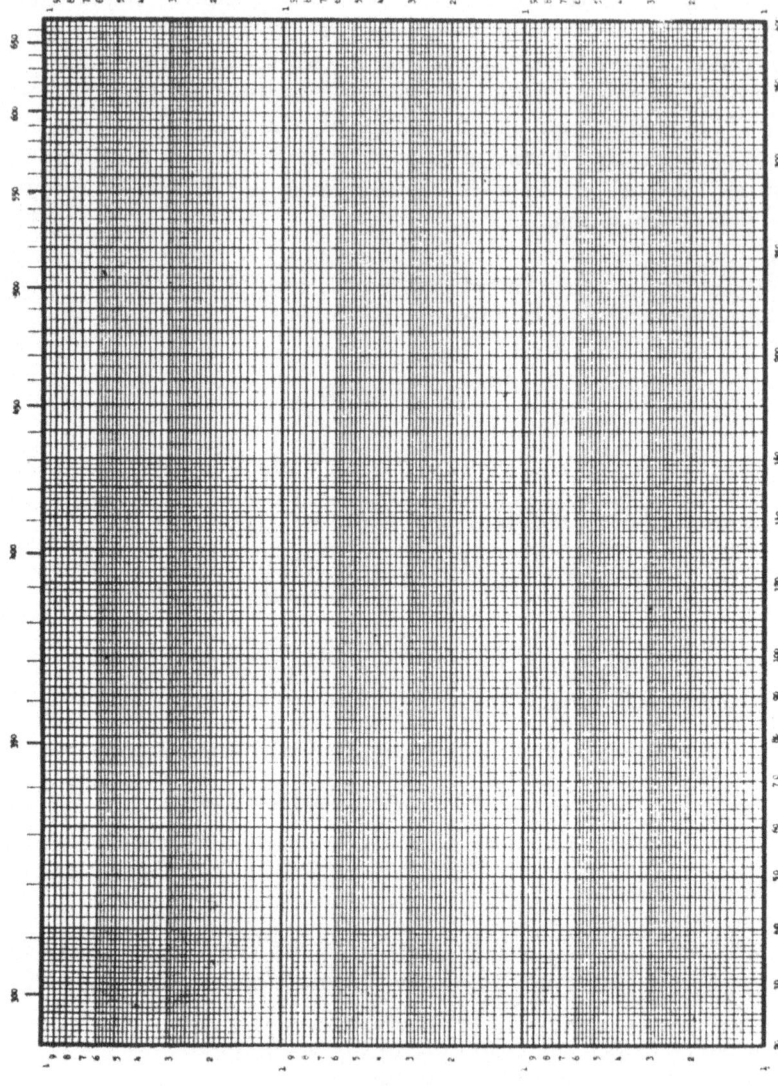

Fig. 4-2. Temperature reciprocal scale. Reproduced by permission of Technical and Engineering Aids for Management, Box 25, Tamworth, N.H. 03886 U.S.A

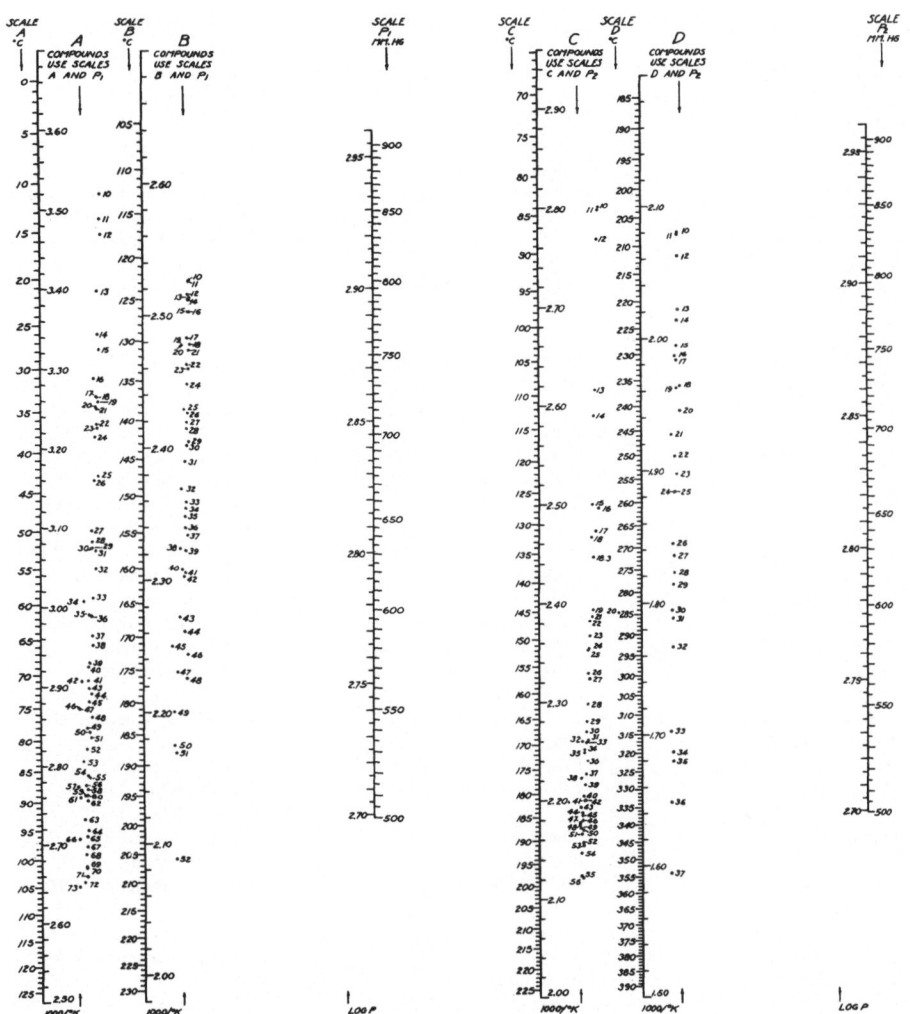

Fig. 4-3. Germann charts for calculating boiling points.

extend them to lower pressures. The following gives a practical illustration of their use.

Example 4-3: What is the boiling point of *o*-nitrotoluene at 510 torr? This compound is represented by point D-13. A straight edge on the pressure scale to the right at 510 torr, through point D-13, intersects scale D at 207°.

Note that our first two examples could also have been solved with this chart, yielding 23.5° as the boiling point of ether under the conditions given.

n-Propyl n-propionate, B-12	Tetrachloroethylene, A-71	Urethane, B-45
n-Propyl n-valerate, B-33	Tetranitromethane, B-1.5	
Pseudocumene, C-27	Toluene, C-13	n-Valeric acid, B-47
Pyridine, C-14	m-Toluidine, C-49	
	o-Toluidine, C-44	Water, A-61
Quinoline, D-13	p-Toluidine, C-45	
	1,1,1-Trichloroethane, A-38	m-Xylene, C-18
1,1,1,2-Tetrachloroethane, B-18	1,1,2-Trichloroethane, A-68	o-Xylene, C-18.3
sym-Tetrachloroethane, B-29	Trichloroethylene, A-48	p-Xylene, C-17

Other pure compounds can be added to Table 4-1 and Fig. 4-3 quite easily. The following example shows just how this is done.

Example 4-4: 1,2-dichlorobenzene is a ring compound that is not included in Germann's original table. Its normal boiling point is 179° and its vapour pressure is 400 mm at 155.8°. Using scale c and the pressure scale to the right (since this is a ring compound), draw two straight lines connecting 760 mm with 179° and 400 mm with 155.8°. The intersection is a point near D29. Label this new point 29a, and add 1,2-dichlorobenzene, D29a, to the table of compounds, Table 4-1. It would be preferable to determine such a point with more than two lines, as Germann did in his original article.

3. Othmer's Modification

Othmer (9) used the Clausius–Clapyron equation to develop a different, versatile method of finding vapour pressures. His method is applicable to a much wider range of pressures, and in addition it is useful for solutions of non-volatile substances.

Othmer's method uses the ratio of equations (4.4) and (4.5), applied to two separate substances, to yield the following equation:

$$\frac{dP_1/P_1}{dP_2/P_2} = \frac{d \ln P_1}{d \ln P_2} = \frac{d \log P_1}{d \log P_2} = \frac{L_1}{L_2}. \tag{4.7}$$

If subscript 1 refers to water and 2 refers to an aqueous solution the vapour pressure of the solution can be predicted nicely. L_1 will be almost identical with L_2, its difference being equal to the heat of solution. In other cases, L_1 and L_2 may not be equal, but both will be nearly constant; their ratio will be even more constant since variations will probably be in the same direction.

Therefore, the graph of (4.7) is a straight line to a quite high degree of precision. Normally L_1/L_2 is regarded simply as a constant, since their separate values are known. In the case of solutions compared to the solvent or other similar solutions, the ratio may be given a value of unity. If

55

possible, it should be evaluated from two points, but this is not normally possible.

Since the Germann chart is adequate for pure liquids, the Othmer method may be illustrated for a solution. Example 4-5 shows how it can be applied graphically to what would otherwise be a fairly complex practical problem.

Example 4-5: A 27.55% aqueous solution of NaOH, saturated with respect to NaCl, has a boiling point of 112.10° at 694.2 torr (8). What would be its boiling point at 464 torr?

Assume that its constant $L_1/L_2 = 1$, since it is an aqueous solution to be compared with water. At 694.2 torr, water boils at 97.47° and at 464 torr it boils at 86.74°.

Let us proceed step by step:

1. Select a log-log graph paper that is 1×1.2 or 1×2 cycles. Use the portion shown in the illustration (Fig. 4-4).

2. Label the left and bottom axes as shown.

3. Draw the solid diagonal as shown. This represents the vapour pressure of water on both axes.

4. Using the table of vapour pressure of water, given here as Table 4-2 (11), locate the boiling point of water at various pressures, as shown. For example, at 103°, the vapour pressure of liquid water is 845 torr. Mark 103° at 845 on the top of the graph.

The graph is now ready for use. The following steps should be followed to solve this problem.

5. Using the temperature scale at the top and the pressure scale to the left, locate the point 112.1°, 694.2 torr.

7. Through this point draw a line parallel to the diagonal. This is the dashed line in the figure.

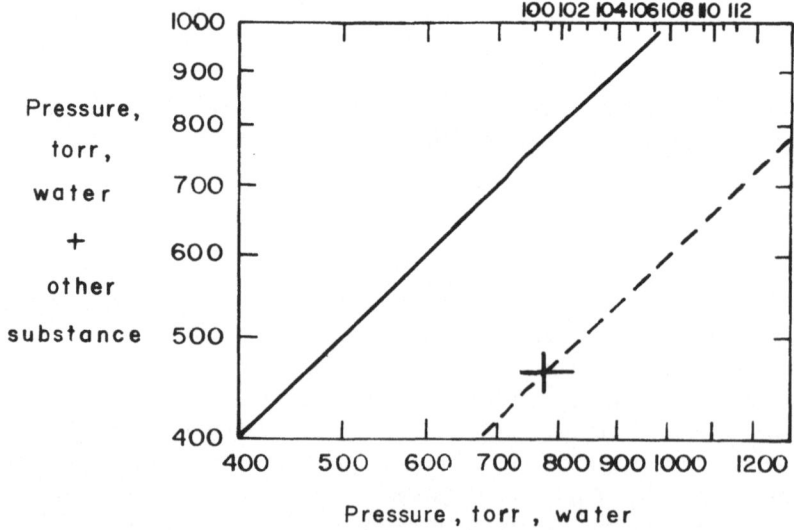

Fig. 4-4. Graphical solution of the boiling point aqueous NaOH saturated with respect to NaCl.

56

Table 4.2. Vapour pressure of water below 100°C. Pressure of aqueous vapour over water in mm of Hg for temperatures from −15.8 to 100°C. Values for fractional degrees between 50 and 89 were obtained by interpolation. (With permission from Robert C. West, 1973–74).

Temp. °C	0.0	0.2	0.4	0.6	0.8	Temp. °C	0.0	0.2	0.4	0.6	0.8
−15	1.436	1.414	1.390	1.368	1.345	42	61.50	62.14	62.80	63.46	64.12
−14	1.560	1.534	1.511	1.485	1.460	43	64.80	65.48	66.16	66.86	67.56
−13	1.691	1.665	1.637	1.611	1.585	44	68.26	68.97	69.69	70.41	71.14
−12	1.834	1.804	1.776	1.748	1.720	45	71.88	72.62	73.36	74.12	74.88
−11	1.987	1.955	1.924	1.893	1.863	46	75.65	76.43	77.21	78.00	78.80
−10	2.149	2.116	2.084	2.050	2.018	47	79.60	80.41	81.23	82.05	82.87
−9	2.326	2.289	2.254	2.219	2.184	48	83.71	84.56	85.42	86.28	87.14
−8	2.514	2.475	2.437	2.399	2.362	49	88.02	88.90	89.79	90.69	91.59
−7	2.715	2.674	2.633	2.593	2.553	50	92.51	93.5	94.4	95.3	96.3
−6	2.931	2.887	2.843	2.800	2.757	51	97.20	98.2	99.1	100.1	101.1
−5	3.163	3.115	3.069	3.022	2.976	52	102.09	103.1	104.1	105.1	106.2
−4	3.410	3.359	3.309	3.259	3.211	53	107.20	108.2	109.3	110.4	111.4
−3	3.673	3.620	3.567	3.514	3.461	54	112.51	113.6	114.7	115.8	116.9
−2	3.956	3.898	3.841	3.785	3.730	55	118.04	119.1	120.3	121.5	122.6
−1	4.258	4.196	4.135	4.075	4.016	56	123.80	125.0	126.2	127.4	128.6
−0	4.579	4.513	4.448	4.385	4.320	57	129.82	131.0	132.3	133.5	134.7
0	4.579	4.647	4.715	4.785	4.855	58	136.08	137.3	138.5	139.9	141.2
1	4.926	4.998	5.070	5.144	5.219	59	142.60	143.9	145.2	146.6	148.0
2	5.294	5.370	5.447	5.525	5.605	60	149.38	150.7	152.1	153.5	155.0
3	5.685	5.766	5.848	5.931	6.015	61	156.43	157.8	159.3	160.8	162.3
4	6.101	6.187	6.274	6.363	6.453	62	163.77	165.2	166.8	168.3	169.8
5	6.543	6.635	6.728	6.822	6.917	63	171.38	172.9	174.5	176.1	177.7
6	7.013	7.111	7.209	7.309	7.411	64	179.31	180.9	182.5	184.2	185.8
7	7.513	7.617	7.722	7.828	7.936	65	187.54	189.2	190.9	192.6	194.3
8	8.045	8.155	8.267	8.380	8.494	66	196.09	197.8	199.5	201.3	203.1
9	8.609	8.727	8.845	8.965	9.086	67	204.96	206.8	208.6	210.5	212.3
10	9.209	9.333	9.458	9.585	9.714	68	214.17	216.0	218.0	219.9	221.8
11	9.844	9.976	10.109	10.244	10.380	69	223.73	225.7	227.7	229.7	231.7
12	10.518	10.658	10.799	10.941	11.085	70	223.7	235.7	237.7	239.7	241.8
13	11.231	11.379	11.528	11.680	11.833	71	243.9	246.0	248.2	250.3	252.4
14	11.987	12.144	12.302	12.462	12.624	72	254.6	256.8	259.0	261.2	263.4
15	12.788	12.953	13.121	13.290	13.461	73	265.7	268.0	270.2	272.6	274.8
16	13.634	13.809	13.987	14.166	14.347	74	277.2	279.4	281.8	284.2	286.6
17	14.530	14.715	14.903	15.092	15.284	75	289.1	291.5	294.0	296.4	298.8
18	15.477	15.673	15.871	16.071	16.272	76	301.4	303.8	306.4	308.9	311.4
19	16.477	16.685	16.894	17.105	17.319	77	314.1	316.6	319.2	322.0	324.6
20	17.535	17.753	17.974	18.197	18.422	78	327.3	330.0	332.8	335.6	338.2
21	18.650	18.880	19.113	19.349	19.587	79	341.0	343.8	346.6	349.4	352.2
22	19.827	20.070	20.316	20.565	20.815	80	355.1	358.0	361.0	363.8	366.8
23	21.068	21.324	21.583	21.845	22.110	81	369.7	372.6	375.6	378.8	381.8
24	22.377	22.648	22.922	23.198	23.476	82	384.9	388.0	391.2	394.4	397.4
25	23.756	24.039	24.326	24.617	24.912	83	400.6	403.8	407.0	410.2	413.6
26	25.209	25.509	25.812	26.117	26.426	84	416.8	420.2	423.6	426.8	430.2
27	26.739	27.055	27.374	27.696	28.021	85	433.6	437.0	440.4	444.0	447.5
28	28.349	28.680	29.015	29.354	29.697	86	450.9	454.4	458.0	461.6	465.2
29	30.043	30.392	30.745	31.102	31.461	87	468.7	472.4	476.0	479.8	483.4
30	31.824	32.191	32.561	32.934	33.312	88	487.1	491.0	494.7	498.5	502.2
31	33.695	34.082	34.471	34.864	35.261	89	506.1	510.0	513.9	517.8	521.8
32	35.663	36.068	36.477	36.891	37.308	90	525.76	529.77	533.80	537.86	541.95
33	37.729	38.155	38.584	39.018	39.457	91	546.05	550.18	554.35	558.53	562.75
34	39.898	40.344	40.796	41.251	41.710	92	566.99	571.26	575.55	579.87	584.22
35	42.175	42.644	43.117	43.595	44.078	93	588.60	593.00	597.43	601.89	606.38
36	44.563	45.054	45.549	46.050	46.556	94	610.90	615.44	620.01	624.61	629.24
37	47.067	47.582	48.102	48.627	49.157	95	633.90	638.59	643.30	648.05	652.82
38	49.692	50.231	50.774	51.323	51.879	96	657.62	662.45	667.31	672.20	677.12
39	52.442	53.009	53.580	54.156	54.737	97	682.07	687.04	717.56	722.75	727.98
40	55.324	55.91	56.51	57.11	57.72	98	707.27	712.40	743.85	749.20	754.58
41	58.34	58.96	59.58	60.22	60.86	99	733.24	738.53	743.85	749.20	754.58
						100	760.00	765.45	770.93	776.44	782.00
						101	787.57	793.18	798.82	804.50	810.21

8. Locate the intersection of 464 torr on the left scale and the new line. Read the temperature from the top scale. In this case, the boiling point is 100.5° according to this graph (the observed value is 100.96°).

4. Myers' Charts

Another useful variation combines Trouton's rule with the Clausius–Clapeyron equation, resulting in nomographs that are quite simple to construct and use (6). Fig. 4-5 furnishes a copy; it could be amplified to extend to the lower pressures needed in the highlands. Example 4-6 shows how this chart is used.

Example 4-6: Using data from Example 4-1, calculate the boiling point of diethyl ether at 600 torr.
Since ether is a normal rather than an associated liquid, lay a straight edge on 35° at the left and 600 torr on the middle scale, and read T on the far right scale, 6.4°. Therefore, the boiling point should be $34.6 - 6.4 = 28°$.

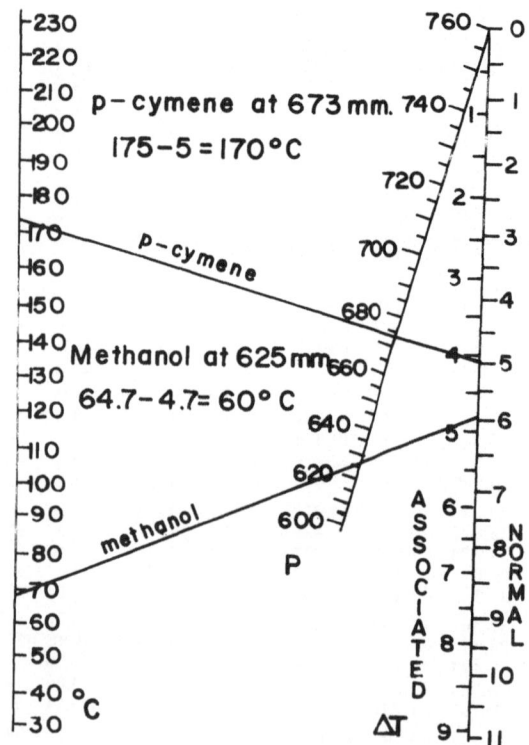

Nomograph for Boiling-Point Correction to 760 mm.

Fig. 4-5. Myer's chart.

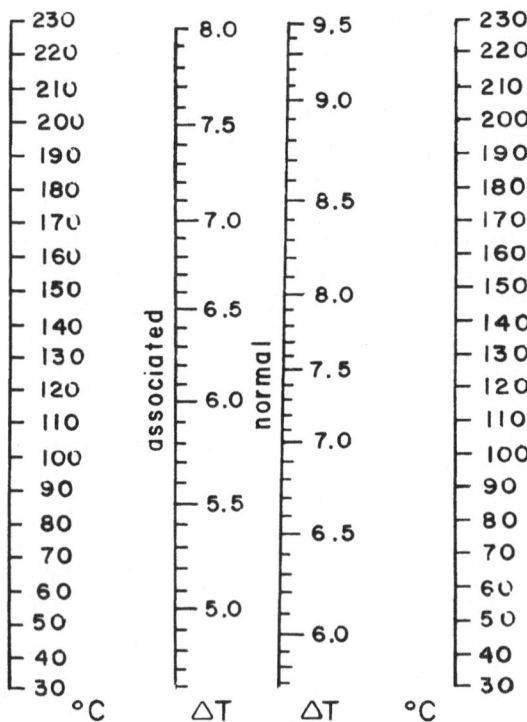

Fig. 4-6. Myers' chart for 617 Torr.

Figure 4-6 illustrates another fine suggestion from the same article. Myers shows that a chart as this one may be constructed for a highlands laboratory for routine use. It should, however, be noted that it is less precise than most other formulations.

5. Dühring's Rule

Dühring (2) presented another of the many useful empirical relations which have proven useful over the years. Even though it was first presented a long time ago, it is still useful.

Dühring's rule has many forms, all of which depend on the approximation that vapour curves meet at a common value for $1/T = 0$. To the extent that this is true, an unknown vapour pressure at any temperature can be estimated from its value at any other temperature.

Since ambient pressures at the earth's surface are not extreme as pressures go, the approximation is ordinarily quite good, though admittedly not as good as many more complex formulations.

Perhaps the greatest advantage of Dühring's rule lies its versatility. Because it requires only one point, it can be applied to any material. It can even be applied to completely unknown materials, as example 4-6 shows.

The normal Dühring plot involves the common logarithms of the vapour

59

pressure in millimetres as ordinate, with reciprocal absolute temperature, usually multiplied by 1000, as abscissa. For low molecular weight inorganic and organic substances without large dipole moments (in particular, without hydroxyl groups) the convergent point is $\log P_0 = 7.6$. For many organic substances, generally of a higher molecular weight, the point is more nearly 8.3.

Examples 4.6: An unknown organic liquid boils at 120° at an ambient pressure of 500 mm Hg. What is its normal boiling point?

Construct a graph of $\log P$ versus $1000/T$ as shown in Fig. 4-7. Draw a

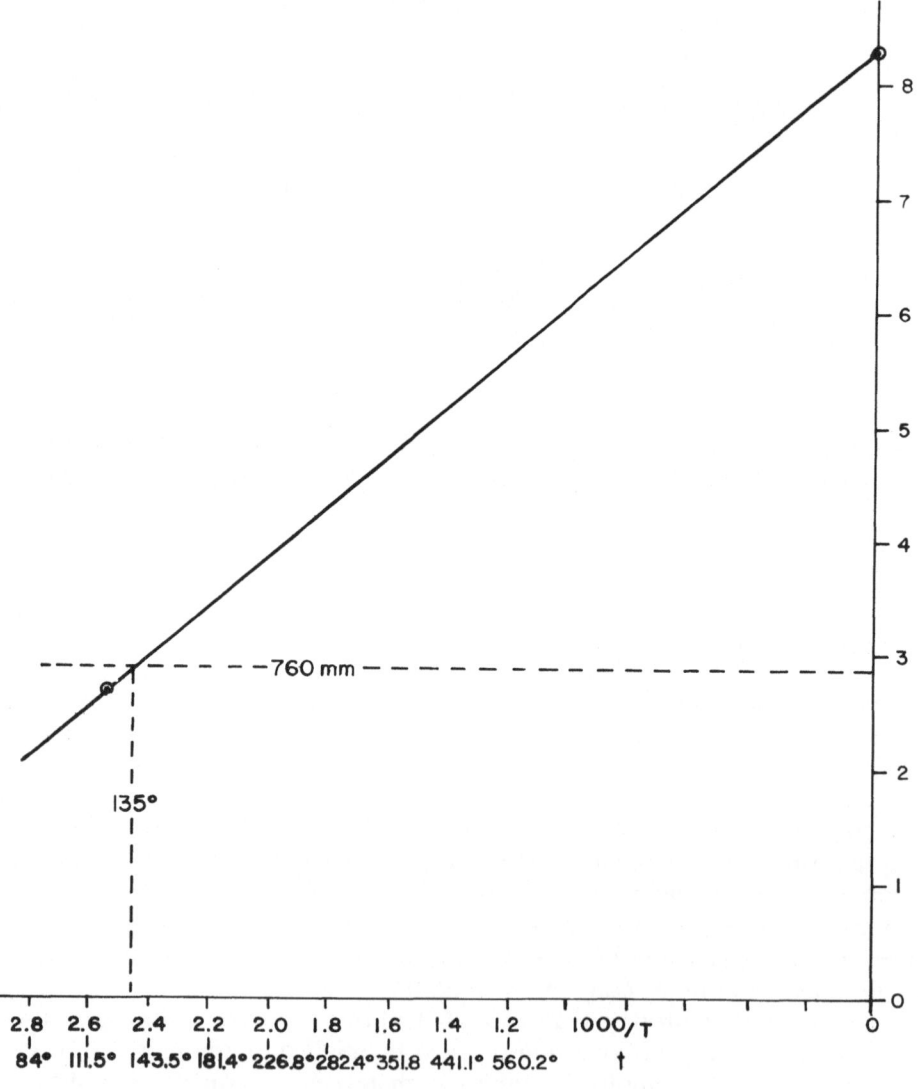

Fig. 4-7. Application of Dühring's rule in example 4-6.

line through the two points: 8.3 at $1000/T = 0$, and $\log 500 = 2.70$ at $1000/T = 1000/(120 + 273.2) = 2.543$.

At $\log 760 = 2.882$, $1000/T$ has a value of 2.46; therefore, $T = 406$ and $t = 133°$, which would be the normal boiling point of this substance. If the convergent point had been taken as 7.6 instead of 8.3, the result would have been $135°$.

Fromherz (3) gives specific instructions for the construction of this type of plot for practical cases, including convenient dimensions for an appropriate graph paper. His excellent book also contains a practical discussion of vapour pressure calculations.

It should also be mentioned here that Dühring's rule, though normally used graphically, is especially well adapted to calculation with pocket electronic calculators designed for scientists. In particular, the calculators programmed to determine the parameters of straight lines are very useful.

For example, with the SR-51 calculator of the Texas Instruments Company, one can enter 8.3, 0, 2.70 and 2.543 to establish the line in Example 4-6. Then he can enter 2.882 and receive, as answer, 2.46. All conversions to logarithms and reciprocals can also be performed on the calculator, mostly during the routine entry of numbers. Given such a calculator, this calculation need not be made graphically.

6. Other Methods

There are a number of other treatments of the vapour pressure problem. Although the methods given above should be adequate for most ambient pressures and ordinary chemical materials, many other methods are also available.

7. Automatic Methods

No account of computational methods would be complete without more mention of automatic and semi-automatic methods of computation. The tremendous growth of large and small computers has motivated many chemists to automate these calculations, both on a large and on a small scale. In addition, although the calculations illustrated in example 4-2 cannot be performed with the average mechanical slide rule, they are completely feasible with the electronic calculators designed for scientists.

Certainly, many vapour pressure calculations are well adapted to this new generation of electronic calculators. In addition, for larger computers, very simple programs can be prepared in FORTRAN or other computer languages to ease vapour pressure-temperature calculations. As a small example, Krieger and Wallace (5) present a useful FORTRAN program for converting handbook vapour pressure to a useful form.

8. References

1. Dalton, J. 1801. *Mem. philos. lit. Soc. Manchester*, 5: 550.
2. Dühring, 1878. Neue Grundgesetze zur rationelle Physik und Chemie. Leipzig.
3. Fromhertz, H. 1960. Physikalisch-chemisches Rechnen in Wissenschaft und Technik. 2nd Ed. Verlag Chemie Weinheim. (Available in English as Physical-Chemical calculations in Science and Industry. Butterworths London, 1964.)
4. Germann, F. E. E. & O. S. Knight. 1934. Line coordinate charts for vapor pressure-temperature data. *Industr. Engin. Chem.*, 26: 467–470.
5. Krieger, A. G. & C. K. Wallace. 1971. Computer Iteration of handbook data. *J. chem. Educ.*, 48: 457.
6. Myers, R. T. 1957. A nomograph for correction of boiling points. *J. chem. Educ.*, 34: 58–59.
7. Monard, C. C. 1929. Correlation of freezing points and vapor pressures of aqueous solutions by Dühring's rule. *Industr. Engin. Chem.*, 21: 139–142.
8. Monard, C. C. & W. L. Badger. 1929. Boiling points of electrolytic solutions. *Industr. Engin. Chem.*, 21: 40–42.
9. Othmer, D. F. 1940. Correlating vapor pressure and latent heat data – A new plot. *Industr. Engin. Chem.*, 32: 841–856.
10. Perry, J. H. Chemical Engineers Handbook. Third Ed. pp. 563–565.
11. Robert, E. W. 1973–1974. Handbook of Chemistry and Physics. Cleveland CRC Press.

Compensating for chemical effects of the lower atmospheric pressure

L. E. Giddings

By accident of geography, laboratory chemistry has mostly developed near sea level. Laboratory procedures have been tailored to low altitude conditions, and little attention has been paid to the problems of the bench chemist in high altitude laboratories.

Barometric pressure is as low as 450 torr in some high altitude metropolitan areas, and the resulting lowering of boiling point of all liquids and solutions is quite noticeable. The effect is seen in all laboratory operations which control reaction temperature by boiling the solutions. This includes reductions and other conversions at the boil in analytical chemistry (3, 4), as well as reflux operations in organic chemistry. Chemists at sea level can ordinarily ignore the ambient pressure, but the high altitude chemist does so at considerable peril.

Many standard methods of analysis, organic syntheses, and other laboratory operations must be altered to compensate for the ambient pressures of the highlands. This chapter considers the simplest method, the extension of the time of reaction.

1. A Simple Model

If a laboratory procedure calls for boiling a solution for a certain period of time, it uses the solution as a temperature control device. If boiling is under total reflux, solvent is not lost and the temperature of boiling usually remains about constant. If there is no reflux arrangement, the temperature will rise, with an overall effect of programming of time and temperature of reaction.

The rate constant of most reactions can be adequately approximated by the empirical Arrhenius equation. In this chapter we assume that the rate constant of all reactions can be represented by the standard expression,

$$k = A \exp(-E/RT), \tag{5.1}$$

in which E is the Arrhenius, or empirical, activation energy.

We further assume that the degree of reaction is a function of kt, where t is the time of reaction, such that a given degree of reaction can be duplicated by any choice of k and t which yields the same product. That is, for a given degree of completion of a reaction, $k_1 t_1 = k_2 t_2$.

This implies that we can compensate a lowering of the reaction rate by extending the time of boiling. Let t_0 and k_0 represent the boiling conditions which yield a given conversion at sea level, as perhaps given in standard analytical methods or standard instructions for organic synthesis. If a lowering in temperature of boiling results in a lower rate constant, k, the boiling time must be increased to t. The factor representing this increase in time, t/t_0, can also be represented by k_0/k:

$$\frac{t}{t_0} = \frac{k_0}{k} = \frac{A \exp(-E/RT_0)}{A \exp(-E/RT)} = \exp\left(\frac{E}{R}\left(\frac{1}{T} - \frac{1}{T_0}\right)\right) \tag{5.2}$$

It remains to evaluate the temperatures of boiling at sea level and at higher altitudes. The change of vapour pressure of a liquid or solution with temperature is conveniently given by the Clausius–Clapeyron equation in an integrated form:

$$\ln P = -\frac{1}{RT} + \text{constant} \tag{5.3}$$

in which L is the heat of vaporization at the normal boiling point, which will be considered constant over the range of temperatures of interest. In practice this equation has been amply proven to be sufficiently accurate for the present purposes.

By combining the previous two equations, the following relation is produced:

$$\frac{t}{t_0} = \left(\frac{P_0}{P}\right)^{E/L}. \tag{5.4}$$

In practice, the factor t/t_0 is multiplied by the time of boiling specified in a given sea level procedure.

Equation (5.4) has the form of the Boltzmann law applied in an elementary way to two separate processes in the same assemblage, and it could be derived on this basis. It reflects the fact that the heat of vaporization is essentially an activation energy, in that all molecules with energy greater than L (on a molecular basis) will evaporate. Although this equation is adequate for the present purpose it must be considered as approximate since neither E nor L is truly independent of temperature, nor can the Boltzmann law be applied in such a simple way.

In general, chemical reactions with higher values of E would be expected to be more sensitive to boiling temperature, since the rate of change of k with temperature is a direct function of E. At the same time, since the rate of change of boiling temperature with pressure is an inverse function of L, liquids of low values of L are more sensitive to the lowering of pressure. Equation (5.4) combines both of these qualitative effects in a simple formula.

2. Practical Correction of Boiling Time

The application of equation (5.4) normally yields conservative boiling times, that is, greater lengths of boiling than necessary. It is in the nature of chemists to specify ample boiling or reaction times, especially when the actual time needed is relatively short. Few analytical or synthetic organic

64

procedures are studied in depth in a kinetic way, especially if reaction times are expected to be short. As a result, ample boiling times at sea level multiplied by t/t_0 yields even more ample boiling times.

If the pressure, P, in a given laboratory is fairly well known, P_0 can be taken to be 760 torr or 1013 mb. It is more probable that altitude is known but pressure is not. In such a case, pressures from Table 5-1 might be conveniently used. These are estimates of the extreme low pressures that might be expected in the absence of extreme weather conditions, such as tornadoes or hurricanes. Since the slope of the low extreme pressure is somewhat smaller than that of the standard pressure as a function of altitude, it will yield a more conservative correction factor.

If E stands for the reaction, L for the solution and the appropriate pressures are known, a simple slide rule calculation will yield t/t_0. This value need only be multiplied by the time of boiling specified for the sea level reaction. However, E and L are normally not known. The following sections are devoted to some practical considerations for the calculation of t/t_0.

3. Arrhenius Activation Energy

An examination of collected kinetic data (11) confirms that the median activation energy of reactions in solution is about 16 kcal. This corresponds to the traditional rule that reaction rates about double for each ten degree increase in temperature. At the same time it can easily be verified that over 90% of the activation energies lie below 25 kcal, and virtually all lie between 8 and 33 kcal.

For certain cases, energies of activation are known *a priori* with great precision. For dissolving metals in strong acids, E is about 5 kcal under diffusion control and from 10 to 20 kcal under chemical control (2).

4. Heat of Vaporization

For aqueous solutions, L can normally be taken as 9.7 kcal, the accepted value for water.

For organic liquids, tabulated valves can often be found. Care must be taken to use L at the boiling point instead of L for absolute zero; as a general rule,

$$L_{\text{boiling point}} = \tfrac{7}{8} \times L_{\text{absolute zero}}.$$
(5.5)(6)

The most volatile solvent normally used in the organic laboratory is diethyl ether, with a normal boiling point of 34° and heat of vaporization of 6.2 kcal. This may be taken as a minimum value for L for volatile solvents since Trouton's rule errs mostly on the positive side. A more average value for solvents boiling at 80° or above would be 7.4 kcal.

One would seldom encounter values of L below about 5.8 kcal, the value for acetic acid. In this case Trouton's constant is abnormally low because the evaporated molecule is a dimer. At the same time, virtually all reactions will have 33 kcal or less for activation energies. These two values define the most extreme case in Table 5-1.

5. Discussion of the Correction Factor

As can be seen from the table, correction for aqueous solutions is normally quite practical. For the great majority of high altitude laboratories, those located below 3000 m, doubling or tripling of the time of boiling would suffice. However, the problem grows quite serious at 4000 m, and it is nearly insurmountable at much higher altitudes. It may be best to consider other ways of compensation at such high altitudes.

Since many organic solvents are not highly associated, their heats of vaporization are often considerably lower than that of water. Below 2000 m the correction is reasonable, but by 3000 m the factor becomes very large. At 4000 m some alternative solution may well be needed. Especially striking is the 26-fold increase of time needed at 4000 m for unfavorable organic chemical reactions.

6. Amenable Reactions

For the equation (5.4) to be valid, an overall reaction rate must have an Arrhenius temperature dependence. More specifically, the following must adequately represent it:

$$f(x) \, dx = A e^{-E/RT} \, dt, \qquad (5.6)$$

in which x represents the concentration of a component of interest, T and t are variables, and A and E are considered constants.

Equation (5.6) can be integrated as follows for the sea level case:

$$\int_{x_1}^{x_2} f(x) \, dx = \int_0^t A \exp\left(-E/RT\right)_{\text{sea level}} dt. \qquad (5.7)$$

For the highlands,

$$\int_{x_1}^{x_2} f(x) \, dx = \int_0^{kt} A \exp\left(-E/RT\right)_{\text{highlands}} dt, \qquad (5.8)$$

in which k is the multiplying factor of time that equalizes the total degree of reaction.

The left members of equations (5.7) and (5.8) are identical, and therefore their right members can be equated and simplified:

$$\int_0^{kt} A \exp\left(-E/RT\right)_{\text{highlands}} dt = \int_0^t A \exp\left(-E/RT\right)_{\text{sea level}} dt. \qquad (5.9)$$

(If A and T are constant, and if T is represented by the Clausius–Clapeyron equation, equation (5.4) follows directly.)

Table 5-1. Correction factors for the highlands, t/t_0. Listed pressures are low extremes that might be expected for normal weather conditions, excluding phenomena such as hurricanes.

	AQUEOUS SOLUTIONS			ORGANIC SOLUTIONS			
				Average Cases			
	Normal case	Most cases	Almost all cases	Normal solvents	Volatile solvents	Most cases	Almost all cases
E, kcal/mol:	16	25	33	16	16	25	33
L, kcal/mol:	9.7	9.7	9.7	7.4	6.2	7.4	5.8
E/L:	1.65	2.57	3,40	2,16	2,58	3,38	5,69
HEIGHT:							
Sea Level ($P_0 = 978.0$ mb)	1.0	1.0	1.0	1.0	1.0	1.0	1.0
1000 m ($P = 852.1$ mb)	1.2	1.4	1.6	1.3	1.4	1.6	2.2
2000 m ($P = 740.1$ mb)	1.6	2.0	2.6	1.8	2.1	2.6	4.9
3000 m ($P = 640.7$ mb)	2.0	3.0	4.2	2.5	3.0	4.2	11
4000 m ($P = 552.3$ mb)	2.6	4.3	7.0	3.4	4.4	6.9	26
5000 m ($P = 473.8$ mb)	3.3	6.4	12	4.8	6.5	12	65
6000 m ($P = 404.4$ mb)	4.3	9.7	20	6.7	9.8	20	152

To be amenable to equation (5.4), then, the reaction's rate equation must be separable into a function of concentration and the Arrhenius term, as in (5.6). This is a satisfactory approximation for the overwhelming majority of cases.

Obviously, 'anti-Arrhenius' reactions are not amenable, since they cannot be represented in this way. Frost and Pearson list several such reactions (1). In explosions, for example, there may be a sudden rise of temperature at an ignition temperature. In catalytic hydrogenations, enzyme reactions, certain forms of the oxidation of carbon, and the nitric oxide-oxygen reaction, the rate can actually decrease with an increase in temperature. It is obvious that all such reactions are excluded.

It should be noted that the order of a reaction is not important, nor is the simplicity of its rate equation. It is merely necessary that the equation be separable in the manner indicated.

7. Solutions under Reflux

A primary assumption in the derivation of (5.4) was that the boiling point of the solution would remain about constant. This in turn implied that the composition remained constant, since almost any change in composition would cause a change in boiling point.

It is, however, obvious that the compositions of solutions must change. Otherwise (5.4) would be meaningless, since we are studying chemical change.

In the typical reflux process the effect may indeed be small. Since the lowering of boiling point by a solute is a colligative property, a reaction in which reactants give products in a 1:1 ratio would result in an insensible change in properties. Production of more moles of product than reactants could conceivably lead to a lowering of the boiling point.

We assume, of course, that solvent is not consumed in the reaction, that solutions are not too concentrated, and that products are not escaping as incondensable vapors. Violations of any of these conditions might cause quite a significant change in the boiling point at reflux.

8. Open Boiling Solutions

Boiling points are not expected to be quite constant in the typical sea level method. In the course of a normal reaction the temperature rises as the composition changes, which results in a programming of reaction temperatures. Indeed, most sea level methods combine a programming of temperature and composition because of evaporation and reaction.

For example, with an involatile solute, the boiling point must increase as the solvent disappears. The limit occurs at saturation, which is not reached in the typical case. Rarely would the temperature rise ten degrees, even to saturation, and in the normal case it would be nearer four or five (8). The boiling point elevation would be less at higher altitudes and lower pressures. However, problems could arise in the concentration of very hygroscopic materials or with solvents of very high molecular weight.

Since we limit ourselves to practical reactions at the boil, we may assume that changes in temperature will seldom pass fifteen or twenty degrees, even with organic reactions in which solvent is consumed. Since programming at high altitudes would roughly parallel sea level conditions, there should be little impediment to the use of (5.4).

Chemical intuition suggests that the factor t/t_0 will be conservative. It is derived for conditions of constant temperature, whereas temperature increases with time. A given temperature increase will have a greater effect at sea level than in the highlands since it occurs at a higher temperature.

This was checked by the numerical integration of equation (5.9) for several practical cases. For example, if the temperature remained constant during a typical reaction at 3000 m, t/t_0 would have a value of 1.8, as predicted by (5.4). For the same general conditions, if the temperature rose linearly, the same factor would have a value of 1.7. If it rose asymptotically to a limit, it would be 1.6. Clearly, (5.4) gives slightly high values and can be freely used.

9. The Egg and Other Special Cases

It is common knowledge that cooking in the highlands can be quite different from cooking in the lowlands. At first glance it would seem possible to apply the factors for organic reactions (from Table 5.1) to cooking times at the boil, but this turns out not to be true.

Consider the case of the egg. Moore (7), in the second edition of his undergraduate text, recalled that to hard-boil an egg at 91° would require twelve hours whereas the same egg could be cooked to the same degree at 100° in ten minutes. Its empirical activation energy is, therefore, about 130 kcal, which would imply a boiling time of nine months at 6000 metres. It is also known that in general, heat denaturation of protein increases 600 times for ten degrees elevation of temperature, a still more extreme case (9).

A study of sterilization times reveals that these are no isolated cases. The accepted moist heat sterilization times for 'most bacterial spores' yield an activation energy of 65 kcal (9). The most famous case, the pasteurization of milk, has an apparent activation energy of about 105 kcal (12). Even the temperature coefficients of sterilization by chemical agents tend to be much higher than the chemist' general rule of a two-fold increase in reaction rate per ten degrees rise in temperature (10).

Clearly the simple factors of the table cannot be applied to the types of organic materials encountered in living systems.

10. Other Ways to Compensate for Altitude

It may not always be feasible to compensate for high altitude by increasing the reaction time. In routine analyses this might unduly extend the time of analysis and increase unit costs, and for various reasons it might also be inappropriate for teaching laboratories.

In heterogeneous reaction, the reaction rate is directly proportional to the exposed area. It may well suffice to augment the surface area by the factor calculated by (5.4). This was shown to be effective in the case of tin analyses (3), but in any case there is a practical limitation to the area that could be exposed.

In some cases, column reductions might be investigated. These have certain efficiencies that might be useful.

A hard boil might be used instead of the traditional gentle boil. Boiling is now quite well understood. In spite of this, there are few data of practical use to the laboratory chemist in this regard.

It may well be possible to increase the concentration of reactants. As a first approximation, the concentration of one reactant might be increased by the factor derived as (5.4). Experimentation would normally be necessary before actual work.

In the case of organic reactions, more drastic measures may be needed, since the effect of altitude is so much more pronounced. In many cases it would be possible to substitute a homologous high-boiling solvent for the specified low boiling solvent, to exchange a higher value of L for the lower value of the specified solvent.

In certain cases, it may be possible to conduct reactions under increased pressure. The standard types of pressure apparatus, such as autoclaves and bombs, would be cumbersome. Nevertheless, devices such as the still described by Levy and Proaño might be useful; their device was used for the determination of normal boiling points at high altitudes (5).

11. References

1. Frost, A. A. & A. G. Pearson. 1961. Kinetics and Mechanism. New York: John Wiley.
2. Gatos, H. C. 1959. The surface chemistry of metals and semiconductors. New York: John Wiley.
3. Giddings, L. E. & G. Schätzle. 1970. Análisis yodimétrico del estaño en la altura I. Rollos de níquél, clavos de Hierro y hoja de alumíno como reductores. *Afinidad*, 27: 533–539; III. La cinetica practica de la reduccion con níquél. *ibid.*, 28.
4. Giddings, L. E. & C. Urioste. 1972. La quimica en la altura. I. Prolongacion de ebullicion en compensacion de la altura. *Afinidad*, 29: 741–745.
5. Levy, L. W. & O. E. Proano. 1957. *J. chem. Educ.*, 34: 440.
6. Moelwyn-Hughes, E. A. 1964. Physical Chemistry. London: Pergamon Press.
7. Moore, W. J. 1955. Physical chemistry. NJ. Prentice-Hall.
8. N.R.C. 1928. International critical tables of numeric data. Vol. 3, pp. 321–329 NY McGraw-Hill Book Co.
9. Sykes, G. 1965. Disinfection and sterilization. Philadelphia: Lippincott. pp. 113–115.
10. Tilley, F. W. 1942. *J. Bact.*, 43: 521.
11. US National Bureau of Standards, 1956. Tables of chemical kinetics. Circ. 510. Suppl. 1. US Government Printing Office, Washington DC.
12. Wilson, G. W. 1942. The Pasteurization of milk. London: E. Arnold.

Barometry

L. E. Giddings

A practical manual for the highlands would be incomplete without a few words on the measurement of pressure. This chapter presents a brief orientation on the measurement of ambient pressure in the highlands laboratory.

1. The Problem

Precision manometry is a legitimate concern to highlands laboratories, and is considerably more important there than in the lowlands. In the absence of actual measurements, one is reduced to speculations and gross estimates.

Without an actual barometer in the laboratory, it is usually necessary to infer the ambient pressure from a knowledge, or supposed knowledge, of altitude. Normally a person is dependent on the reported altitude of an adjacent airport. But in point of fact, the altitudes of most isolated airports have probably been inferred from readings of aneroid barometers, which have, in turn, been calibrated in terms of some standard atmosphere. Reported altitudes of airports are sometimes nothing more than crude estimates.

Even in areas that cannot be considered isolated, the matter can be complex. One might expect to use information from a reporting weather station. But unless the station's technicians are more sophisticated than usual, this source can also be troublesome.

Consider, for example, a reporting station which uses only an aneroid barometer. Consider at the same time the following brief discussion of pressure (6):

> "In meteorology, the pressure measured at the observation station is called the '*actual pressure*', but a standard station elevation is used, so that a small correction is necessary for the observed pressure to obtain to (*sic*) '*station pressure*'. This in turn is regularly reduced to '*sea-level pressure*', for ready correlation on synoptic charts; in high areas (e.g., Rocky Mts.), another common datum, e.g., 5000 feet, may be used."

The situation is more complex yet, since the World Meteorological Organization has not yet suggested a uniform scheme for highland areas.

Even if the reporting system were unambiguous, it does not seem wise to trust the calibration of aneroid barometers of isolated areas. According to the WMO, such barometers need to be compared with a mercury barometer at least once a week (8), and certainly few stations have access to a properly calibrated and installed mercury barometer.

Even assuming correct measurements of properly calibrated instruments, it can be risky using reported pressure measurements without detailed knowledge of how they are obtained. Meteorologists are faced with the

71

problem of interpreting pressures synoptically, that is to say, simultaneously over a large region. To them, altitudes of reporting stations are only a distraction which bias their data. As a result, the World Meteorological Organization recommends that for low level stations, the following value be added to all measurements of pressure (8):

$$C = 34.68 \frac{H_P}{T_V} \text{ millibars},$$

where H_P is the height of the station in metres above sea level, and T_V is the absolute virtual air temperature on the Kelvin scale. The WMO also describes other methods that can be used to reduce pressures to a common standard (9). The virtual temperature can be approximated by the actual air temperature.

2. Standards of the WMO

For the practical problems of meteorology the WMO has defined several classes of precision barometers, among which the following are of greatest interest to us:

A_r A barometer of category A which has been selected by regional agreement as a reference standard for barometers of that region.
B_r A barometer of category B in a region, which the national meteorological services of the region agree to use as the standard barometer of the region, in the event that a barometer of category A is unavailable in the region.
S 'A good station barometer', a mercury barometer, usually of the fixed cistern (Kew pattern) or a Fortin type barometer.

For details of class A, B and many others, see references (8). Several different kinds are needed for the complex business of standardizing station barometers, which are described in the same document.

A good station barometer needs to be standardized against a regional standard. The regional standard barometers currently in existence are listed here, by region.

I	Pretoria, South Africa	A_r
II	Calcutta, India	B_r
III	Rio de Janeiro, Brazil	A_r
IV	Buenos Aires, Argentina	B_r
	Maracay, Venezuela	B_r
IV	Washington, D.C. (Gaithersburg Maryland), U.S.A.	A_r
V	Melbourne, Australia	A_r
VI	London, United Kingdom	A_r
	Leningrad, U.S.S.R.	A_r

Paris, France A_r
Hamburg, Federal Republic
of Germany A_r

According to the WMO, the main requirements of a good station barometer are the following:

(a) Its accuracy should not vary over long periods of time
(b) It should be easy and quick to read
(c) It should be transportable without loss of accuracy
(d) The bore of the tube should not be less than 8 mm and should preferably be 9 mm
(e) The tube should be prepared and filled under vaccum
(f) The actual temperature for which the scale is assumed to give true readings (at standard gravity) should be engraved on the barometer; the scales should preferably be calibrated to give correct readings at 0°C
(g) The meniscus should not be flat
(h) In calibration against a standard barometer whose index errors are known and allowed for, the following tolerances for a station barometer should not be exceeded:
Maximum permissible error at about
1000 mb..±0.3 mb
Maximum permissible error at any other pressure for a barometer whose range:
Does not extend below 800 mb ....±0.5 mb
Extends below 800 mb ....±0.8 mb
Difference between errors over an interval of 100 mb or less   0.3 mb

Further, the WMO suggests that remedial action is generally required when the absolute error exceeds ±0.010 inch or ±0.3 mb: The change or error exceeds +0.003 inch or ±0.1 mb; the error undergoes sudden shifts or exhibits erratic behaviour; or the instrument shows mechanical defects. Remedial action consists of repeating the comparison with a standard and repairing or replacing the barometer.

The same WMO document states that aneroid barometers are less reliable, but it lists the chief requirements of 'a good aneroid barometer' as follows:

(a) It should be compensated for temperature so that the reading does not change more than 0.5 mb for a change of temperature of 30°C
(b) The scale errors at any point should not exceed 0.5 mb and should remain within this tolerance over periods of at least a year when in normal use
(c) The hysteresis should be sufficiently small to ensure that the difference in reading before a change of pressure of 50 mb and after return to the original value does not exceed 0.5 mb

(d) It should be capable of withstanding ordinary transit risks without introducing inaccuracies outside the limits specified above

The instrument should always be read in the same position, vertical or horizontal, as it had when being calibrated. It should be tapped lightly before it is read. As far as possible it should be read to the nearest 0.1 mb.

It is obvious that aneroid barometers have distinct advantages for remote and isolated highlands areas. At the same time, most of the requirements for good ones are harder to meet in high areas. The temperature will normally be much lower than the calibrated temperature, and calibration will normally be performed at considerably lower pressure than ambient highlands pressure. Even ordinary transit is likely to be more risky.

The WMO manual, from which these data have been taken would be a valuable acquisition to a laboratory interested in measurements of the environment. In addition to barometry, it contains general information on meteorological stations, instruments, and observations, including temperature, humidity, winds, precipitation, evaporation, radiation, visibility, clouds and many more subjects.

The following units and standard conditions are extracted from the same reference:

1 Millibar (mb) = 0.750062 millimetre of mercury under standard conditions (mm Hg)$_n$ (also called torr)

1 (mm Hg)$_n$ = 1.333224 mb.

Taking one inch to exactly equal 25.4 millimetres, we have the following conversion factors:

1 mb = 0.0295300 inch of mercury under standard conditions (in. Hg)$_n$

1 (in. Hg)$_n$ = 33.8639 mb

1 (mm Hg)$_n$; 0.0393708 (in. Hg)$_n$.

The value of 0°C is the standard temperature to which mercury barometer readings are reduced for the purpose of relating actual density of mercury at its observed temperature to the standard density of mercury at 0°. The standard density of mercury at 0° is taken to be 13.5951 gr/cc and it is regarded as an incomprehensible fluid.

Barometric readings have to be reduced from local acceleration due to gravity to the standard (normal) gravity. The value of standard gravity is regarded as a conventional constant, g = 980.665 cm/sec^2. This is recognized by scientists as a gravity datum to which reported barometric data shall refer, but it does not represent the gravity of latitude 45°, at sea level.

Scales on mercury barometers for meteorological purpose should be so graduated that they yield true pressure readings directly in standard units when the entire instrument is maintained at the standard temperature of 0° and at the standard value of gravity of 980.665 cm/sec.^{2} This implies that the scales of Fortin barometers graduated in mm or inches will yield true linear readings when the scale is maintained at a temperature of 0°.

3. Mercury Barometers

As a practical matter, if a high altitude laboratory is interested in precision barometry, it must purchase a special kind of mercury barometer. Most commercial barometers can only be used to about 1000 m or 3000 feet of altitude. Those for higher altitudes must generally be ordered as a special modification.

In an attempt to locate sources of barometers for high altitudes, one of the authors (LG) wrote to all manufacturers and agents listed in two standard commercial references (7). The few sources of precision mercury barometers that came from this exercise are listed in foot-note below.* Not a single response was received from Europe, perhaps because such barometers are normally custom build. In general, the standard barometers which were found were manufactured only for altitudes to 3000 ft, but several companies indicated that instruments for higher altitudes were available on special order. The authors welcome more information on sources, which will be included in any future editions of this book.

The so-called 'flight barometers' may provide sufficient precision for laboratory uses in the highlands. These are apparently used for calibration of aircraft altimeters at simulated altitudes.

4. Aneroid Barometers

Aneroid barometers will suffice for low precision purposes of an average laboratory, but again, most commercially available models do not register pressures low enough for use in the highlands.†

Recording barographs are essentially adjustable recording aneroid instruments. Again, most sea-level instruments will not give useful readings in the highlands, since their mechanisms cannot be adjusted to their low pressures. The few that may be adaptable to the highlands, judging from published descriptions, are listed in the footnote.

* *Sources of mercury barometers for high altitude:* Science Associates, Inc., 230 Nassau Street, Princeton, New Jersey 08540 USA (model 325(1) to 12000 ft (3700 m), scales in inches, mm and mb): Weather Measure Corporation, Box 41257, Sacramento, California 95841 USA (model B 222 with special modifications for altitudes over 3000 m; The Scott and Fetzer Company, Meriam Instruments Division, 10920 Madison Avenue, Cleveland, Ohio 44102 USA (Meriam Model 31EG10 on special order for scale modification); Kahl Scientific Instrument Corporation, 737 West Main, El Cajon, California USA (Model 09AM600, to 4000 m)

† *Sources of precision aneroid barometers for high altitude:* Science Associates (model 360 and 361 to 30,000 ft, model 363(7) to 6000 ft, model 317 to 27,000 ft, model 306–3 to 9500 ft, and model 309 to 9500 ft); Weather Measure Corporation (model BM60-D for 2000 to 3000 m).
Sources of barographs (self-recording aneroid) for high altitude: Science Associates (model 355, model 351(1): Weather Measure Corporation (models B201, B211, and B231 to 4000 m).

5. Constructing a Mercury Barometer

The principle of the mercury barometer is so exceedingly simple that one is naturally led to the idea of constructing one. Middleton's fine, readable account of the history of such efforts (4) can furnish some valuable perspective, as well as references. He includes sufficient details on all aspects to discourage any idle attempt to construct a precision barometer.

Of course, there are any number of ways of constructing one that has no pretense of high precision. If a truly fine vaccuum system is available in a laboratory (perhaps a mechanical vaccuum pump of high efficiency, followed by a diffusion pump), one might simply evacuate a long tube inserted in a disch of mercury (very slowly, of course), and measure the distance from the pool to the top of the column. But the disadvantage of home-built instruments of all kinds (assuming they are properly built and filled) is just this problem of measurement. Scales must be corrected for variation of temperature, as does the density of the mercury, and there are gravity corrections needed as well. All these corrections are normally built into the scales of commercial instruments. In any case, the appropriate corrections are given in many standard handbooks.

Details of construction of precision barometers are not particularly easy to find. Reports in the technical literature are usually quite old and therefore not well cataloged or abstracted. One report of a barometer that seems feasible is presented in Germann (1). Again, accuracy of readings will depend on the precision of the scale and the technique of measurement. In this case it may be expedient to use a cathetometer and apply temperature corrections. Other corrections are mentioned in the article, along with the accuracy obtained by the author in two instruments which he constructed.

6. The Contrabarometer

An article in the *Scientific American* magazine describes a 'contrabarometer' of simple construction from materials common to laboratories (5). This device traces its ancestry to the Hooke two-level barometer described by Middleton (4), and suffers the severe practical limitations of that instrument. It has the advantage of greater movement of the surface for a given change in pressure, essentially an amplification of the scale.

Despite its disadvantages, vividly detailed by Middleton, this device is mentioned here for several reasons. Details on its construction are readily available, and its construction is apparently simple enough to recommend it to amateurs with average skill in manipulating glass. But more important, adapting its design to the lower pressures of the highlands will be an interesting exercise for students. Sufficient theoretical and practical details for both purposes are included in the reference.

7. Electronic Devices

The mercury barometer is certainly the instrument of choice for most laboratory uses, and precision aneroid barometers are quite useful if their limitations are understood. Nevertheless, electronic devices are now available and may well be useful for some purposes. As might be expected, the easiest ones to use seem to be the most expensive, and the least expensive ones seem to be the most difficult to use. One should understand the catalog description of one of these instruments perfectly before ordering it – several require external devices before they can be used.*

8. The Hypsometer

An alternative to other types of barometers is the hyposometer. In this device, the boiling point of a liquid is used to determine the ambient pressure, and this pressure is converted to altitude by using a standard atmosphere.

Hypsometers were quite in vogue with mountain climbers in the 1800's, which saw the invention of ingenious portable devices. Still, in their simple forms, they have been virtually abandoned. Although their principle is simple, the realization of that principle is, again, just not that easy. Measuring the boiling point of a liquid is probably not an appropriate task for a simple, portable system In any case, its use as an altimeter depends on a standard atmosphere, and so it can have no advantage over modern aneroid barometers and altimeters based on them.

The name hypsometer is derived from its use as an altitude-measuring device. Thus, it is badly named for a pressure-measuring device. Still, in the sense of a self-enclosed device for measuring pressures, it may still have some advantages, especially since 'height' for a meteorologist is actually a pressure measurement in upper air. Hypsometers still find use in meteorology, especially in balloon flights, and particularly because their precision increases rapidly at the low pressure of modern weather balloon altitudes.

The practical range of these devices can be increased by a pressure-subtracting devices, as shown in Fig. 6-1. These diagrams are taken from an article which describes, in detail, the construction of a pressure-subtracting hypsometer with electronic controls (3). The device appears to be capable of high precision in sea-level or high-altitude laboratories. It would probably be expedient to modify the author's design of the measuring and control circuits in light of modern integrated electronic devices.

9. A 'Chemical' Method

In the absence of more appropriate devices, it is apparent that a well-equipped chemical or physics laboratory will still be able to determine the ambient pressure with a known accuracy. The measurement of boiling

---

* Source of some electronic devices for pressure measurement for use at altitude: Science Associates (models 360A, 360B, and 360C digital barometers for different ranges, usuable as received; model 363A pressure transducer with 363B converter, requires recorder or other output device, specifiable to 4000 metres; etc). Because of advances in electronics, the state of instrumentation is in flux. It is best to contact manufacturers directly.

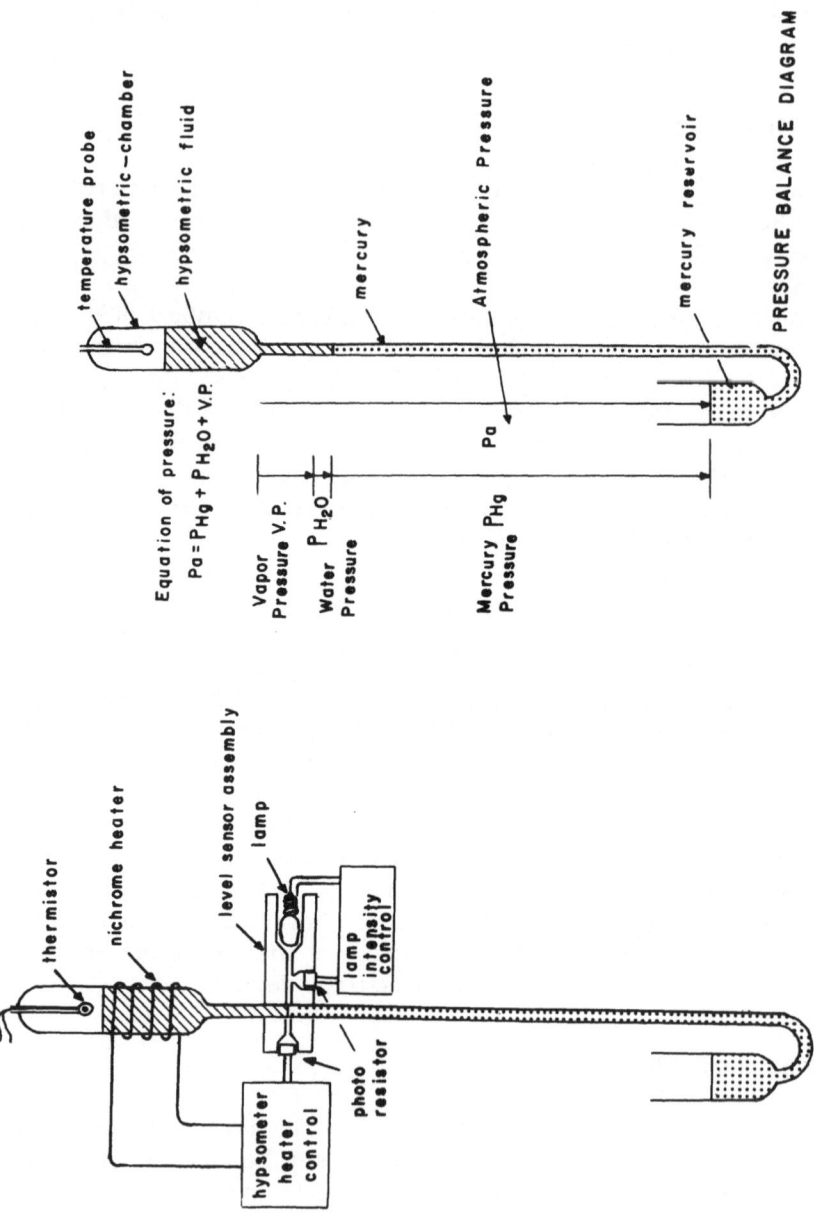

Fig. 6-1. A pressure-subtracting hypsometer.

temperature of liquids allows us to calculate the ambient pressure, provided that the temperature can be measured with sufficient accuracy. An open reflux system, familiar to all chemists, should suffice. Table 4-2, which presents the vapour pressure of water as a function of temperature, will allow the direct conversion of boiling temperature to ambient pressure.

One must be conservative. 1°C of error in measurement at an altitude of 3000 m would cause an error of 20 mm or 27 mb in pressure, equivalent of 300, nearly 100 feet, of altitude. This measurement involves absolute accuracy, not precision. It is not easy to ensure the absolute calibration at the boiling point of water. Indeed, the accuracy of a thermometer is more commonly judged by its reading of boiling points, given that the ambient pressure is known.

10. References

1. Germann, Albert F., 1915. A Modified Precision Barometer, *J. Amer. Chem. Soc.*, 36: 2456–2562.
2. Handbook of Type A-1 Barometers/Manometers (flight barometer for calibration of aircraft altimeters), with supplement, Hass Instrument Corporation, Washington, D.C.
3. Lichfield, Ernest W. 1959. The Pressure-Subtracting Hypsometer: a New Instrument for Measuring Atmospheric Pressure, *J. appl. Meteorol.*, 8: 975–980.
4. Middleton, Knowles, W. E. 1964. 'The History of the Barometer'. Johns Hopkins, Baltimore.
5. Muehlner, E. reported by C. L. Stong, 1971. The Amateur Scientist, *Scientific American*, 225 (1): 111–114.
6. Rhodes W. Fairbridge, 1967. 'The Encyclopedia of Atmospheric Sciences and Astrogeology', 782, Reinhold, N.Y.
7. Thomas Register, Thomas Publishing Co., N.Y. (1977); Jaegers Europa Register, Darmstadt (1975).
8. World Meteorological Organization, 'Guide to Meteorological Instruments and Observing Practices', 4th ed. sect. 3.4, Geneva (1971) (WMO-no. 8. TP. 3) (WMO publications are available from regional distributors; in the United States, the UNIPUB organization, New York, is the distributor).
9. World Meteorological Organization, Notes on the standardization of pressure reduction methods in the international network of synoptic stations, Technical Note No. 61, Geneva (1964) (WMO-No. 154. TP. 74).

I Psychrometry

L. E. Giddings

The highlands scientist often needs to measure the atmospheric humidity, and his instrument of choice will likely be a sling psychrometer. However, the use of this instrument in the highlands has traditionally been hindered by one problem above all others: the lack of psychrometric tables for atmospheric pressures corresponding to high surface altitudes (1, 2).

Although there is international agreement on the definition of relative humidity, the practical problem facing the laboratory scientist has been serious. If we assume that he works in a laboratory at 4200 m (13,800 feet) with an average barometric pressure of 450 torr (600 mb, 17.7 in Hg), and that he has constructed a sling psychrometer and operated it, he must somehow convert his readings to values of relative humidity.

Table 7-1 shows his practical problem. The several columns correspond to psychrometric tables that are available to him. The German tables yield columns *c* and *d* after a complex calculation of deviation from sea level values; these are probably the most exact conversions available, even though they were specifically designed for the semi-automatic self-ventilated Assman psychrometer. Column *e* presents results from the direct reading Rechard tables for Fahrenheit temperatures; however, the pressure does not correspond to his altitude, and the tables are not generally available. Much more accessible are the remaining tables, but they are all for lower altitudes and for Fahrenheit temperatures.

This chapter presents a complete set of psychrometic tables for the highlands. Tables 7-2 through 7-9 were constructed on a metric basis, consistent with the Marvin (3) and Rechard (4) tables. Below freezing point, calculations are based on supercooled water. Calculations below $-5°$ were omitted since their usefulness and precision seem greatly in doubt.

Above the freezing point the tables are probably accurate to a percentage point or two, although great care and cleanliness of equipment are required for such accuracy. Below the freezing point they should not be considered more precise than about 5 per cent. Experimental errors are high in any case, and the author suggest great conservatism in the use of all psychrometric charts.

It should be emphasized that these tables are not definitive. Instead they are designed only to be consistent with some existent tables, to serve the practical needs of technical people until definitive tables become available.

1. Alternatives to the Psychrometer

Since there is little hope of great precision in ordinary psychrometry, the highlands scientist might well find that other methods of measuring atmospheric water better serve his needs. A wide variety of methods is practical.

81

Table 7-1. Relative humidity at 4200 m according to several psychrometric tables. Values in parentheses were not included in the tables in this chapter. Asterisks identify values which were missing from the author's copy of reference 2.

Air temperature °C	Lowering of the temperature °C	a Giddings (2) 450 Tor Water °C	b Giddings (2) 450 Tor Ice °C	c German (1) Water °C	d German (1) Ice °C	e Rechard (4) 19 inches Water °F	f Canada 2500 feet Ice °F	g Marvin (3) 23 inches Ice °F	h Marvin (3) 29 inches Ice °F	i German (1) sea level Water °C	j German (1) sea level Ice °C	k Canada (5) sea level Ice °F
−20	0.5	(80)	(76)	—	63	75	58	71	65	—	55	56
	1.0	(61)	(54)	—	46	51	34	42	30	—	29	29
	1.5	(42)	(31)	—	29	26	11	14	—	—	3	—
	2.0	(23)	(8)	—	—	2	—	—	—	—	—	—
−10	1.0	(79)	(76)	—	71	78	65	72	67	—	63	63
	2.0	(58)	(54)	—	*	51	41	45	34	—	*	36
	3.0	(38)	(32)	—	*	28	15	18	3	—	*	9
−5	1.0	83	(82)	73	79	81	75	80	76	77	74	74
	2.0	67	(65)	67	64	63	56	59	52	54	53	53
	3.0	52	(48)	51	50	46	38	39	29	32	33	33
	4.0	36	(32)	35	35	29	19	20	4	10	13	13
0	1.25	83	(82)	83	82	82	80	79	77	78	78	78
	2.5	67	(65)	67	68	64	61	50	55	56	58	58
	5.0	37	(31)	36	38	31	24	24	13	14	19	19
	7.5	9	(1)	—	—	—	—	—	—	—	—	—
10	2.5	76		76		76	73	74	71	71	—	71
	5.0	55		55		54	48	50	44	44	—	45
	7.5	35		35		34	25	28	20	19	—	20
	10.0	18		—		15	—	8	—	—	—	—

Table 7-2. Psychrometric table for an atmospheric pressure of 650 torr or 867 millibars, equivalent to an altitude of about 1300 metres. Tables 7-2 through 7-9 present values of relative humidity corresponding to measurements taken from sling psychrometers. Air temperature, that is to say, the dry bulb temperature, is found at the left of the tables. The lowering of the temperature, that is, the difference of the readings of the two thermometers, is found above each column. For example, at an altitude of 4200 metres (Table 7-6), if the air temperature is 5° and the dry bulb thermometer reads 3°, the relative humidity would be 78%.

	0.2	0.4	0.6	0.8	1.0	1.2	1.4	1.6	1.8	2.0	2.2	2.4	2.6	2.8	3.0	3.2	3.4	3.6	3.8	4.0	4.2	4.4	4.6	4.8	5.0	5.5	6.0	6.5	7.0	7.5	8.0	8.5	9.0	9.5	10.0	10.5	11.0
−5	96	92	87	83	79	75	71	67	63	59	55	51	47	43	39	35	31	28	24	20	16	12	9	5	1												
−4	96	92	88	84	80	76	72	69	65	61	57	53	50	46	42	39	35	31	28	24	20	17	13	10	6												
−3	96	92	89	85	81	77	74	70	66	63	59	56	52	49	45	42	38	35	31	28	24	21	17	14	11	2											
−2	96	93	89	86	82	79	75	72	68	65	61	58	54	51	48	44	41	38	34	31	28	25	21	18	15	7											
−1	97	93	90	86	83	79	76	72	69	66	62	59	56	53	50	47	43	40	37	34	31	26	23	20	13	11	3										
0	97	93	90	87	84	80	77	74	70	68	64	61	58	55	52	49	46	43	40	37	34	29	26	23	16	12	4										
1	97	94	91	87	84	81	78	75	72	69	66	63	60	57	54	51	48	45	43	40	37	34	31	28	23	19	12	8	1								
2	97	94	91	88	85	82	79	76	73	70	67	64	62	59	56	53	51	48	45	42	40	37	34	32	26	23	16	10	3								
3	97	94	92	88	86	82	80	77	74	72	69	67	64	61	58	55	53	50	47	45	42	40	37	35	29	26	20	13	7	3							
4	97	94	92	89	87	84	81	78	76	73	71	69	66	63	60	57	55	52	49	47	45	42	40	37	31	29	23	17	11	6	4						
5	97	95	92	89	87	84	82	80	77	75	73	70	67	64	61	59	56	54	51	49	47	44	42	40	34	31	26	20	15	9	8	1					
6	97	95	92	90	88	85	82	80	78	75	73	71	69	66	63	60	58	56	53	51	49	46	44	42	36	34	29	23	18	13	11	6	1				
7	97	95	93	90	88	86	83	80	78	76	73	71	69	67	64	62	60	57	55	53	50	48	46	44	39	36	31	26	21	16	14	10	5				
8	98	95	93	90	88	86	83	81	79	77	74	72	70	67	65	63	61	59	56	54	52	50	48	46	44	39	34	29	24	19	14	13	10	5			
9	98	95	93	91	88	86	84	82	79	77	75	73	71	69	67	65	63	61	59	56	54	52	50	48	46	41	36	31	27	22	17	16	11	6	3		
10	98	95	93	91	88	87	84	82	80	78	76	74	72	70	68	66	64	62	60	58	56	54	53	51	49	43	38	33	29	24	20	19	14	9	6	2	
11	98	96	93	91	89	87	85	83	81	79	77	75	73	71	69	67	65	63	61	59	57	55	54	53	51	45	40	36	31	27	23	21	18	12	9	6	2
12	98	96	94	92	90	88	86	84	82	80	78	76	74	72	70	68	66	64	62	60	58	56	54	53	51	46	42	38	33	29	25	24	20	15	12	9	5
13	98	96	94	92	90	88	86	84	82	80	79	77	75	73	71	69	67	65	63	62	60	58	56	54	52	48	44	40	35	31	27	26	22	17	14	12	8
14	98	96	94	92	90	88	87	85	84	82	80	78	76	74	72	70	68	66	64	62	61	59	57	55	53	49	45	41	37	33	30	28	24	20	17	14	11
15	98	96	94	92	90	89	87	85	84	82	80	79	77	75	73	71	69	67	65	63	62	60	58	57	55	51	47	43	39	35	32	30	26	21	18	16	14
16	98	96	94	92	91	89	87	85	83	81	80	78	76	74	73	71	69	68	66	64	63	61	59	58	56	52	48	45	41	37	34	30	26	23	20	16	13

	0.5	1.0	1.5	2.0	2.5	3.0	3.5	4.0	4.5	5.0	5.5	6.0	6.5	7.0	7.5	8.0	8.5	9.0	9.5	10.0	10.5	11.0	11.5	12.0	12.5	13.0	13.5	14.0	14.5	15.0	15.5	16.0	17.0
18	95	91	87	82	78	74	70	66	62	58	55	51	47	44	40	37	34	30	27	24	21	18	15	12	9	6	3						
20	96	91	87	83	79	75	71	68	64	60	57	53	50	47	43	40	37	34	31	28	25	22	19	16	13	11	8	5	3	1			
22	96	92	86	84	80	76	73	69	66	62	59	55	52	49	46	43	40	37	34	31	28	26	24	20	17	15	12	10	7	5	3	1	
24	96	92	85	81	77	74	71	67	64	61	58	54	51	48	46	43	40	37	34	32	29	26	24	21	19	16	14	11	9	7	5	3	1
26	96	92	85	81	77	75	72	68	65	62	59	56	53	50	47	44	42	39	36	34	31	28	26	24	22	19	16	14	11	9	7	5	1
28	96	93	86	82	79	76	73	70	67	64	61	58	55	52	49	46	44	41	38	36	34	31	28	26	24	21	19	16	14	12	10	5	5
30	96	93	86	83	80	77	74	71	68	65	62	59	56	54	51	48	46	43	41	38	36	34	31	28	26	24	21	19	18	14	12	7	9
32	97	93	87	84	81	78	75	72	69	66	63	60	58	55	53	50	48	45	43	41	39	36	34	31	29	27	24	22	20	17	15	10	12
34	97	93	87	84	82	79	76	73	70	67	64	61	59	57	54	52	49	47	45	43	41	38	36	34	31	29	27	25	23	21	18	14	15
36	97	94	88	85	82	80	77	74	71	69	66	64	61	59	57	54	52	50	48	46	43	42	38	36	34	31	29	27	26	24	22	17	18
38	97	94	88	86	83	80	78	75	73	70	68	66	63	61	59	57	55	53	51	48	46	44	42	40	38	36	34	31	29	27	26	24	22
40	97	94	89	86	84	81	79	76	74	72	69	67	65	63	61	58	56	54	52	50	48	46	44	42	40	38	36	34	31	29	28	26	24
42	97	94	89	87	84	82	80	77	75	73	70	68	66	64	62	59	57	55	53	51	49	47	45	43	41	39	36	34	33	31	29	28	26
44	97	94	91	89	86	83	80	78	76	73	71	69	67	65	63	61	59	57	54	52	50	48	46	44	41	39	36	34	33	31	30	28	28

Table 7-3. Psychrometric table for an atmospheric pressure of 600 torr or 800 millibars, equivalent to an altitude of about 1950 metres.

	0.2	0.4	0.6	0.8	1.0	1.2	1.4	1.6	1.8	2.0	2.2	2.4	2.6	2.8	3.0	3.2	3.4	3.6	3.8	4.0	4.2	4.4	4.6	4.8	5.0	5.2	5.4	5.6	5.8	6.0	6.2	6.4	6.6	6.8	7.0	7.2	7.4	7.6
-5	96	92	88	84	80	76	73	69	65	61	57	53	50	46	42	39	35	31	28	24	20	17	13	10	6	3												
-4	96	92	89	85	81	78	74	70	67	63	59	56	52	49	45	42	38	35	31	28	24	21	18	14	11	8	4	1										
-3	96	93	89	86	82	79	75	72	68	65	61	58	54	51	48	44	41	38	34	31	28	25	21	18	15	12	9	6	3									
-2	97	93	90	86	83	80	76	73	70	66	63	60	57	53	50	47	44	40	37	34	31	28	25	22	19	16	13	10	7	4	1							
-1	97	93	90	87	83	80	77	74	71	68	65	62	58	55	52	49	46	43	40	37	34	31	29	26	23	20	17	14	11	9	6	3						
0	97	94	91	88	84	81	78	75	72	69	66	63	60	57	54	51	49	46	43	40	37	35	32	29	26	23	21	18	15	13	10	7	5	2				
1	97	94	91	88	85	82	79	76	73	70	68	65	62	59	56	54	51	48	45	43	40	37	35	32	29	27	24	21	19	16	14	11	9	6	4	1		
2	97	94	92	89	85	83	79	77	75	72	69	67	63	61	58	55	52	50	48	45	42	40	37	35	32	30	27	25	22	20	17	15	13	10	8	5	3	
3	97	94	92	89	86	83	81	78	75	73	70	69	65	62	60	57	55	52	50	47	45	42	40	37	35	32	30	28	25	23	20	18	16	14	12	9	7	5
4	97	95	92	89	87	84	81	78	76	74	71	69	66	64	61	59	56	54	52	49	47	44	42	40	37	35	33	31	28	26	24	22	19	17	15	13	11	9
5	97	95	92	90	87	85	82	80	77	75	72	70	67	65	63	60	58	55	53	51	49	46	44	42	40	38	35	34	31	29	27	25	22	20	18	16	14	12
6	97	95	93	90	88	85	83	80	78	76	73	71	69	66	64	62	59	57	55	53	51	48	46	44	42	40	38	36	34	31	29	27	25	23	21	19	17	15

	0.2	0.4	0.6	0.8	1.0	1.2	1.4	1.6	1.8	2.0	2.2	2.4	2.6	2.8	3.0	3.2	3.4	3.6	3.8	4.0	4.2	4.4	4.6	4.8	5.0	5.2	5.4	5.6	5.8	6.0	6.5	7.0	7.5	8.0	8.5	9.0	9.5	10.0	10.5	11.0
7	98	95	93	90	88	86	83	81	79	76	74	72	67	65	63	61	59	57	54	52	50	48	46	44	44	41	38	36	34	34	29	24	19	15	10	5	1			
8	98	95	93	91	88	86	84	81	79	77	75	72	69	67	64	62	60	58	55	53	52	50	48	46	46	43	41	38	36	36	31	27	22	18	13	9	4	1		
9	98	95	93	91	89	87	84	82	80	78	76	74	70	68	65	63	61	59	57	55	54	51	49	48	49	45	43	41	38	38	34	29	25	20	16	12	8	4		
10	98	96	94	91	89	87	85	83	81	79	77	75	71	69	67	65	63	61	59	57	57	55	53	51	51	48	45	43	41	41	36	31	27	23	19	15	11	8	4	
11	98	96	94	92	89	88	85	84	82	80	77	75	72	71	69	67	65	63	61	59	58	56	54	53	54	50	48	45	43	43	38	34	29	25	21	17	14	10	7	3
12	98	96	94	92	90	88	85	84	82	80	78	76	73	72	70	68	66	64	62	61	59	57	55	54	56	52	50	48	46	46	40	36	32	28	24	20	16	13	9	6
13	98	96	94	92	90	88	86	84	83	81	79	77	74	73	71	69	67	65	63	62	60	58	57	55	58	54	52	50	48	48	44	39	34	30	26	22	19	15	12	9
14	98	96	94	92	90	88	86	84	83	81	79	77	75	74	72	70	68	66	65	63	61	60	58	57	59	55	53	51	49	50	46	41	37	34	30	26	23	19	16	13
15	98	96	94	92	91	89	87	85	83	82	80	78	76	75	73	71	69	67	66	64	62	60	59	57	60	57	55	52	50	51	47	43	39	35	32	28	25	21	18	16
16	98	96	94	93	91	89	87	85	84	82	80	79	76	75	73	72	70	68	67	65	63	62	61	59	61	58	56	54	52	54	50	45	41	37	34	31	27	24	21	18
17	98	96	94	93	91	90	88	86	84	83	81	79	77	76	74	73	71	69	68	66	64	63	62	61	62	59	57	55	53	55	51	46	43	39	36	32	29	26	23	21
18	98	96	95	93	91	90	88	86	85	83	81	80	77	76	74	73	72	70	69	67	65	64	62	62	63	61	58	56	54	56	52	48	44	41	38	34	31	29	25	23
19	98	97	95	93	91	90	88	86	85	84	82	81	78	77	75	74	72	71	69	68	66	65	63	63	64	62	60	57	56	57	53	49	47	43	39	36	34	31	27	25
20	98	97	95	93	92	90	88	87	85	84	82	81	78	77	76	74	73	71	70	68	67	66	64	64	61	59	58	55	53	58	55	51	48	45	42	39	37	34	30	27

	0.5	1.0	1.5	2.0	2.5	3.0	3.5	4.0	4.5	5.0	5.5	6.0	6.5	7.0	7.5	8.0	8.5	9.0	9.5	10.0	10.5	11.0	11.5	12.0	12.5	13.0	13.5	14.0	14.5	15.0	15.5	16.0	16.5	17.0
22	96	92	88	84	81	77	73	70	66	63	60	56	53	50	47	44	41	38	33	30	27	25	22	20	17	15	12	10	8	5	3			
24	96	92	89	85	81	78	74	71	68	64	61	58	55	52	49	46	44	41	35	33	30	28	25	23	21	18	16	14	12	9	7	5	3	1
26	96	93	89	86	82	79	75	72	69	66	63	60	57	54	51	48	46	43	38	35	33	30	28	26	24	21	19	17	15	13	11	9	7	5
28	96	93	89	86	83	79	76	73	70	66	64	61	58	56	53	51	48	45	40	38	35	33	31	29	26	24	22	20	18	16	14	12	11	9
30	97	93	90	86	83	80	76	73	70	68	64	61	58	57	54	52	49	47	42	40	38	35	33	31	29	27	25	23	21	20	18	16	14	12
32	97	93	90	87	84	81	78	74	71	68	65	62	59	57	54	53	50	49	44	42	40	38	35	33	31	30	28	26	24	22	21	19	17	16
34	97	94	90	87	84	81	78	75	72	69	66	64	61	57	57	54	52	50	46	44	42	40	38	36	34	32	30	28	26	25	23	21	20	18
36	97	94	91	88	85	82	79	76	73	71	68	65	62	59	57	55	53	49	47	45	43	41	39	37	35	33	32	30	28	27	25	23	21	19
38	97	94	91	88	85	83	80	76	73	71	68	65	63	59	59	56	54	51	49	47	45	43	41	39	38	36	34	32	31	29	27	25	24	21
40	97	94	91	88	86	83	80	77	74	71	68	67	62	60	60	57	55	52	50	48	47	45	42	40	40	38	35	33	32	30	28	27	25	25
42	97	94	91	89	86	83	80	77	74	72	69	67	65	62	60	58	56	53	51	49	47	45	44	42	40	40	38	35	33	32	30	27	27	27

Table 7-4. Psychrometric table for an atmospheric pressure of 550 torr or 733 millibars, equivalent to an altitude of about 2650 metres.

Band 1 (temperatures −5 to 7 °C) — columns are wet-bulb depressions (°C):

t	0.2	0.4	0.6	0.8	1.0	1.2	1.4	1.6	1.8	2.0	2.2	2.4	2.6	2.8	3.0	3.2	3.4	3.6	3.8	4.0	4.2	4.4	4.6	4.8	5.0	5.2	5.4	5.6	5.8	6.0	6.2	6.4	6.6	6.8	7.0	7.2	7.4	7.6	7.8
−5	96	92	89	85	81	78	74	70	67	63	60	56	52	49	45	42	38	35	32	28	25	21	18	15	11	5	1												
−4	96	93	89	86	82	79	75	72	68	65	61	58	55	51	48	45	41	38	35	32	28	25	22	19	16	12	9	6	3										
−3	97	93	90	86	83	80	76	73	70	66	63	60	57	54	50	47	44	41	38	35	32	29	26	23	20	17	14	11	8	5	2								
−2	97	94	90	87	84	81	77	74	71	68	65	62	59	56	53	50	47	44	41	38	35	32	29	26	23	20	17	15	12	9	6	4	1						
−1	97	94	91	87	84	81	78	75	72	69	66	63	60	58	55	52	49	46	43	40	38	35	32	29	27	24	21	18	16	13	10	8	5	3					
0	97	94	91	88	84	82	79	76	73	71	68	65	62	59	57	54	51	48	46	43	40	38	35	32	30	27	25	22	19	17	14	12	9	7	4	2	1		
1	97	94	91	88	86	83	79	77	74	72	69	66	64	61	58	56	53	50	48	45	43	40	38	35	33	30	28	25	23	20	18	15	13	11	8	6	5	2	
2	97	94	92	89	86	83	81	78	75	73	70	68	65	62	60	57	55	52	50	47	45	43	40	38	35	33	31	28	26	24	21	19	17	14	12	10	9	5	3
3	97	95	92	89	87	84	81	79	76	74	71	69	66	64	61	59	57	54	52	49	47	45	42	40	38	36	33	31	29	26	24	22	20	18	16	13	11	9	7
4	97	95	92	90	87	85	82	79	77	75	72	70	68	65	63	61	58	56	54	51	49	47	45	42	40	38	36	34	31	29	27	25	23	21	19	17	15	13	11
5	97	95	93	90	88	85	82	80	78	76	73	71	69	66	64	62	60	57	55	53	51	49	47	44	42	40	38	36	34	32	30	28	26	24	22	20	18	16	14
6	98	95	93	90	88	86	83	81	79	77	74	72	70	68	65	63	61	59	57	55	53	51	48	46	44	42	40	38	36	34	32	30	28	26	25	23	21	19	17
7	98	95	93	91	88	86	84	82	80	77	75	73	71	69	67	64	62	60	58	56	54	52	50	48	46	44	42	40	38	37	35	33	31	29	27	25	23	22	20

Band 2 (temperatures 8 to 16 °C):

t	0.2	0.4	0.6	0.8	1.0	1.2	1.4	1.6	1.8	2.0	2.2	2.4	2.6	2.8	3.0	3.2	3.4	3.6	3.8	4.0	4.2	4.4	4.6	4.8	5.0	5.5	6.0	6.5	7.0	7.5	8.0	8.5	9.0	9.5	10.0	10.5	11.0	11.5
8	98	95	93	91	89	87	84	82	80	78	76	74	72	70	68	66	64	62	60	58	56	54	52	50	48	43	39	34	30	25	21	17	12	8	4			
9	98	96	93	91	89	87	85	83	81	79	77	75	73	71	69	67	65	63	61	59	57	55	53	51	50	45	41	36	32	28	23	19	15	11	7	4		
10	98	96	94	92	89	87	85	83	81	79	77	75	73	72	70	68	66	64	62	60	58	57	55	53	51	47	43	38	34	30	26	22	18	14	10	7	4	
11	98	96	94	92	90	88	86	84	82	80	78	76	74	72	71	69	67	65	63	61	60	58	56	54	53	48	44	40	36	32	28	24	20	17	13	9	7	2
12	98	96	94	92	90	88	86	84	82	81	79	77	75	73	71	70	68	66	64	63	61	59	57	55	54	50	46	42	38	34	30	26	22	19	16	12	10	8
13	98	96	94	92	90	89	86	85	83	81	80	78	76	74	73	71	69	67	66	64	62	60	58	57	55	51	48	43	40	36	32	28	24	22	18	14	13	11
14	98	96	94	92	90	89	87	85	83	82	80	79	77	75	74	72	70	68	67	65	63	61	59	58	56	53	49	45	41	37	34	30	26	24	20	17	16	14
15	98	96	94	93	91	89	87	86	84	82	81	79	77	76	74	73	71	69	67	65	64	62	60	59	58	54	50	46	43	39	36	32	28	26	22	19	19	17
16	98	96	95	93	91	89	88	86	84	82	82	80	78	77	76	75	73	71	69	68	66	65	63	62	59	55	51	48	44	41	37	34	31	28	25	21	21	20

Band 3 (temperatures 18 to 36 °C):

t	0.5	1.0	1.5	2.0	2.5	3.0	3.5	4.0	4.5	5.0	5.5	6.0	6.5	7.0	7.5	8.0	8.5	9.0	9.5	10.0	10.5	11.0	11.5	12.0	12.5	13.0	13.5	14.0	14.5	15.0	15.5	16.0	16.5	17.0	17.5
18	96	91	87	83	79	75	72	68	64	61	57	54	50	47	44	40	37	34	31	28	25	22	20	17	14	12	9	6	4	1					
20	96	92	88	84	80	76	73	69	66	62	59	56	52	49	46	43	40	37	34	31	29	26	23	21	18	16	13	11	8	6	4	1			
22	96	92	88	85	81	77	74	71	67	64	61	58	54	51	48	45	43	40	37	34	32	29	27	24	22	20	17	15	12	10	8	6	4		
24	96	92	89	85	82	78	75	72	68	65	62	59	56	53	51	48	45	42	40	37	34	32	29	27	25	23	20	18	16	14	12	10	8	4	
26	96	92	89	86	82	78	76	72	68	67	64	61	58	55	53	50	47	44	42	39	37	35	32	30	28	25	23	21	19	17	15	13	11	7	2
28	96	93	90	86	83	79	76	73	70	68	65	62	59	57	55	51	49	46	44	41	39	37	34	32	30	28	26	24	22	20	18	16	14	11	6
30	96	93	90	87	83	80	77	74	71	69	66	63	61	58	57	53	50	48	46	43	41	39	37	35	32	30	28	26	25	23	21	19	17	14	9
32	97	93	90	87	84	80	78	75	72	70	67	64	62	60	58	54	52	50	47	45	43	41	39	37	35	33	31	29	27	25	23	22	20	17	13
34	97	93	91	87	85	81	79	76	73	71	68	65	63	61	59	56	53	51	49	47	45	42	40	38	36	35	33	31	29	27	26	24	22	19	16
36	97	94	91	88	85	82	79	77	74	72	69	66	64	61	61	57	55	52	50	48	46	44	42	40	38	36	34	33	31	29	28	26	24	21	18

Table 7-5. Psychrometric table for an atmospheric pressure of 500 torr or 667 millibars, equivalent to an altitude of about 3400 metres.

0.2	0.4	0.6	0.8	1.0	1.2	1.4	1.6	1.8	2.0	2.2	2.4	2.6	2.8	3.0	3.5	4.0	4.5	5.0	5.5	6.0	6.5	7.0	7.5	8.0	8.5	9.0	9.5	10.0	10.5	11.0	11.5	12.0	12.5	13.0	13.5	14.0
96	93	89	86	87	79	75	72	69	65	62	58	55	52	49	40	32	25	17	9	1																
97	93	90	86	83	80	77	73	70	67	64	60	57	54	51	44	35	28	20	13	6	3															
97	93	90	87	84	81	78	74	71	68	65	62	59	56	53	46	38	32	24	17	10	8	1														
97	94	91	88	85	82	78	76	73	70	67	64	61	58	55	48	41	37	27	21	14	11	5	4													
97	94	91	88	85	82	79	77	74	71	68	65	62	60	57	50	44	40	30	24	18	15	9	7	2												
97	94	91	89	86	83	80	77	75	72	69	67	64	61	59	52	46	42	33	27	21	19	13	11	7	1											
97	94	92	89	86	84	81	78	76	73	70	68	65	63	60	54	48	44	36	30	24	22	17	15	10	5	4										
97	95	92	89	87	84	81	79	77	74	72	69	66	64	62	56	50	46	38	33	27	25	20	18	13	8	7	3									
97	95	92	90	87	85	82	80	77	75	73	70	68	66	63	57	52	48	41	35	30	27	23	21	16	11	10	6	2								
98	95	93	90	87	85	82	81	78	76	74	71	69	67	65	59	53	50	43	38	33	30	25	23	19	15	13	9	5	1							
98	95	93	90	88	86	83	81	78	76	74	72	68	68	65	60	55	52	45	40	35	32	28	26	22	17	16	12	8	4	1						
98	95	93	91	88	86	84	82	80	78	75	73	70	69	67	62	57	53	47	42	37	35	30	28	24	20	19	15	11	8	4	3					
98	96	92	91	89	86	84	82	80	78	75	74	71	69	67	62	58	55	48	44	39	37	32	30	27	23	21	18	14	10	7	6	3				
98	96	93	91	89	87	84	83	80	78	76	74	72	70	68	63	59	57	50	46	41	39	35	33	29	25	23	20	16	13	10	9	6	3			
98	96	93	92	89	87	85	83	81	80	77	76	73	71	69	63	61	59	52	47	43	41	37	35	31	27	26	22	19	16	13	12	9	6	2		
98	96	93	92	89	88	85	84	81	80	77	76	73	72	70	64	62	60	53	49	45	42	38	36	33	29	28	24	21	18	15	14	11	8	5	2	
98	96	94	92	90	88	85	84	82	80	78	76	74	73	71	65	63	61	54	50	46	44	40	38	35	31	30	26	23	20	17	16	13	11	8	5	
98	96	94	92	90	88	86	85	82	81	78	77	74	73	72	66	64	62	56	52	48	45	42	40	36	33	31	28	25	22	19						
98	96	94	92	90	89	86	85	83	82	79	78	75	74	72	67	65	63	57	53	49	47	43	41	38	35											
98	96	94	93	91	89	87	86	83	82	80	79	76	75	73	68	66		58	54	50	48	45														
98	96	95	93	91	89	87	86	84	83	81	79	76	76	74	69	66		59	55	52																

0.5	1.0	1.5	2.0	2.5	3.0	3.5	4.0	4.5	5.0	5.5	6.0	6.5	7.0	7.5	8.0	8.5	9.0	9.5	10.0	10.5	11.0	11.5	12.0	12.5	13.0	13.5	14.0	14.5	15.0	15.5	16.0	16.5	17.0	17.5
95	91	86	82	78	74	70	66	62	58	54	50	47	43	40	36	33	30	26	23	20	17	14	11	8	5	2								
96	91	87	83	79	75	71	67	64	60	56	53	49	46	43	39	36	33	30	27	24	21	18	16	13	10	7	5	2						
96	92	88	84	79	76	72	69	65	62	58	55	52	48	45	42	39	36	33	30	28	25	22	19	17	14	12	9	7	5	2				
96	92	88	84	80	77	74	70	68	63	60	57	54	51	48	45	42	39	36	33	31	28	26	23	21	18	16	13	11	9	7	4	2		
96	92	89	85	81	77	74	71	68	65	62	59	56	53	50	47	44	41	39	36	34	31	29	26	24	21	19	16	15	13	11	8	6	3	3
96	93	89	85	81	78	75	71	69	65	62	59	56	53	50	47	44	41	39	36	34	34	31	29	27	24	22	19	18	16	14	12	10	6	6
96	93	89	86	81	79	75	72	69	66	63	60	57	54	53	49	46	44	41	38	38	36	34	31	30	27	24	22	21	19	17	15	13	9	10
96	93	90	86	82	79	76	73	70	67	64	61	59	56	55	51	48	46	43	41	40	38	36	34	32	30	27	24	23	22	20	18	16	12	13
97	93	90	87	83	80	77	74	71	68	65	63	60	57	56	52	50	47	45	43	42	40	38	36	34	32	29	26	26	24	22	21	20	15	16
97	94	90	87	83	80	77	74	72	68	67	64	61	59	57	54	51	49	47	44	44	42	40	38	36	34	32	29	28	26	24	23	24	17	18
97	94	90	87	83	81	78	75	72	70	67	65	62	60	59	55	53	50	48	46	45	43	41	39	38	36	34	32	30	29	27	25	26	20	20
97	94	91	88	84	81	79	76	73	71	68	66	63	61	59	56	54	52	50	48	47	45	43	41	39	37	36	34	32	30	29	27	26	22	23
97	94	91	88	85	82	80	77	74	72	69	67	64	62	60	57	55	53	51	49	47	45	43	41	39	37	36	34	32	30	29	27	26	24	23

Table 7-6. Psychrometric table for an atmospheric pressure of 450 torr or 600 millibars, equivalent to an altitude of about 4150 metres.

	0.2	0.4	0.6	0.8	1.0	1.2	1.4	1.6	1.8	2.0	2.2	2.4	2.6	2.8	3.0	3.2	3.4	3.6	3.8	4.0	4.2	4.4	4.6	4.8	5.0	5.5	6.0	6.5	7.0	7.5	8.0	8.5	9.0	9.5	10.0	10.5	11.0	11.5
-5	97	93	90	87	83	80	77	74	70	67	64	61	58	55	52	49	45	42	39	36	33	30	28	25	22	15	7											
-4	97	94	90	87	84	81	78	75	72	69	66	63	60	57	54	51	48	45	42	39	36	34	31	28	25	18	11	5										
-3	97	94	91	88	85	82	79	76	73	70	67	64	61	59	56	53	50	47	45	42	39	36	34	31	28	22	15	9	3									
-2	97	94	91	88	85	83	80	77	74	71	68	66	63	60	58	55	52	50	47	44	42	39	37	34	31	25	19	13	7									
-1	97	94	91	89	86	83	80	78	75	72	70	67	64	62	59	57	54	52	49	47	44	42	39	37	34	28	22	16	11	5								
0	97	94	92	89	86	84	81	79	76	73	71	68	66	63	61	58	56	53	51	49	46	44	42	39	37	31	25	20	14	9	4							
1	97	95	92	90	87	84	82	79	77	74	72	70	67	65	62	60	58	55	53	51	48	46	44	41	39	34	28	23	18	12	7	2						
2	97	95	92	90	87	85	83	80	78	75	73	71	68	66	64	61	59	57	55	53	50	48	46	44	41	36	31	26	21	15	11	6	1					
3	98	95	92	90	88	86	83	81	79	76	74	72	69	67	65	63	61	58	56	54	52	50	48	46	44	38	33	28	24	18	14	9	5					
4	98	95	93	91	88	86	84	81	79	77	75	73	70	68	66	64	62	60	58	56	54	52	50	48	46	41	36	31	26	22	17	13	8	4				
5	98	95	93	91	89	86	84	82	80	77	76	74	71	69	67	65	63	61	59	57	55	53	51	49	47	43	38	33	29	24	20	16	11	7	3			
6	98	95	93	91	89	87	85	83	80	78	76	74	72	70	68	66	64	62	60	58	57	55	53	51	49	44	40	35	31	27	23	19	14	10	6			
7	98	96	94	91	89	87	85	83	81	79	77	75	73	71	69	67	65	64	62	60	58	56	54	52	51	46	42	37	33	29	25	22	17	13	9	6	2	
8	98	96	94	92	90	88	86	84	82	80	78	76	74	72	70	68	66	65	63	61	59	57	56	54	52	48	44	39	35	31	28	24	20	16	12	9	5	
9	98	96	94	92	90	88	86	84	82	80	78	77	75	73	71	69	67	66	64	62	60	59	57	55	53	49	45	41	37	33	30	26	22	19	15	12	8	5

	0.5	1.0	1.5	2.0	2.5	3.0	3.5	4.0	4.5	5.0	5.5	6.0	6.5	7.0	7.5	8.0	8.5	9.0	9.5	10.0	10.5	11.0	11.5	12.0	12.5	13.0	13.5	14.0	14.5	15.0	15.5	16.0	16.5
10	95	90	85	81	76	72	67	63	59	55	51	47	43	39	35	32	28	24	21	18	14	11	8	4	1								
12	95	91	86	82	77	73	69	65	61	57	53	50	46	42	39	35	32	29	25	22	19	16	13	10	7	4	1						
14	96	91	87	83	79	75	71	67	63	59	56	52	49	45	42	38	35	32	29	26	23	20	17	14	12	9	6	4	1				
16	96	92	87	83	80	76	72	68	65	61	58	54	51	48	44	41	38	35	32	29	27	24	21	18	16	13	11	8	6	3	1		
18	96	92	88	84	80	77	73	70	66	63	59	56	53	50	47	44	41	38	35	33	30	27	25	22	20	17	15	12	10	8	6	3	1
20	96	92	88	85	81	78	74	71	68	64	61	58	55	52	49	46	43	41	38	35	33	30	28	25	23	21	18	16	14	12	10	8	5
22	96	93	89	85	82	78	75	72	69	66	63	60	57	54	51	48	46	43	40	38	35	33	31	28	26	24	21	19	17	15	13	11	9
24	96	93	89	86	83	79	76	73	70	67	64	61	58	55	53	50	47	45	42	40	38	35	33	31	29	27	24	22	20	18	16	14	13
26	97	93	89	86	83	80	77	74	71	68	65	62	60	57	54	52	49	47	44	42	40	37	35	33	31	29	27	24	23	21	19	17	16
28	97	93	90	87	84	81	78	75	72	69	66	63	61	58	56	53	51	48	46	44	42	39	37	35	33	31	29	27	25	24	21	20	18
30	97	93	90	87	84	81	78	76	73	70	67	65	62	59	57	55	52	50	48	45	43	41	39	37	35	33	31	29	28	26	24	22	21
32	97	94	90	87	85	82	79	76	73	71	68	65	63	61	58	56	54	51	49	47	45	43	41	39	37	35	33	31	30	28	26	25	23
34	97	94	91	88	85	82	79	77	74	71	69	66	64	62	59	57	54	53	50	48	46	44	42	40	39	37	35	33	31	30	28	27	25
36	97	94	91	88	85	83	80	77	75	72	70	67	65	63	60	58	56	54	52	50	48	46	44	42	40	38	37	35	33	32	30	28	27

Table 7-7. Psychrometric table for an atmospheric pressure of 400 torr or 533 millibars, equivalent to an altitude of about 5100 metres.

	0.2	0.4	0.6	0.8	1.0	1.2	1.4	1.6	1.8	2.0	2.2	2.4	2.6	2.8	3.0	3.2	3.4	3.6	3.8	4.0	4.2	4.4	4.6	4.8	5.0	5.5	6.0	6.5	7.0	7.5	8.0	8.5	9.0	9.5	10.0	10.5	11.0	11.5
−5	97	94	91	87	84	81	78	75	72	69	66	63	60	58	55	52	49	46	43	41	38	35	32	30	27	20	13	7	1									
−4	97	94	91	88	85	82	79	76	73	71	68	65	62	59	57	54	51	48	46	43	40	38	35	33	30	24	17	11	5									
−3	97	94	91	88	86	83	80	77	75	72	69	66	64	61	58	56	53	51	48	45	43	40	38	35	33	27	21	15	9	3								
−2	97	94	92	89	86	84	81	78	76	73	70	68	65	63	60	58	55	53	50	48	45	43	40	38	36	30	24	18	13	7	2							
−1	97	95	92	89	87	84	82	79	77	74	71	69	66	64	62	60	57	55	52	50	47	45	43	40	38	32	27	21	16	11	6	1						
0	97	95	93	90	87	85	82	80	78	75	72	70	68	65	63	61	58	57	54	52	49	47	45	43	40	35	30	24	19	14	9	4						
1	97	95	93	90	88	85	83	80	78	75	73	71	69	66	64	62	59	57	55	53	50	49	47	45	43	37	32	27	22	17	13	8	3					
2	98	95	93	90	88	86	83	81	79	76	74	71	70	68	65	63	60	58	55	54	51	49	47	45	43	40	35	30	25	20	16	11	7	2				
3	98	95	93	91	88	86	84	82	80	77	74	72	70	69	67	65	63	60	58	56	53	51	50	48	45	42	37	32	28	23	19	14	10	5	2			
4	98	95	94	91	89	87	84	82	80	77	75	73	71	69	68	66	64	62	60	58	56	54	52	50	48	44	39	34	30	26	21	17	13	9	5	1		
5	98	96	94	91	89	87	85	83	81	78	76	74	72	70	68	67	65	63	61	59	58	55	54	52	51	45	41	36	32	28	24	20	16	12	8	5	1	
6	98	96	94	92	89	87	85	83	82	78	76	74	73	71	69	67	66	64	62	60	59	57	55	53	51	47	43	38	34	30	26	23	19	15	11	8	4	
7	98	96	94	92	90	88	86	84	82	79	77	75	73	72	70	68	67	65	63	61	60	58	57	55	53	49	44	40	36	32	29	25	21	17	14	10	7	4
8	98	96	94	92	90	88	86	84	82	80	77	76	74	73	71	69	68	66	65	63	61	60	59	57	55	51	46	42	38	34	31	27	23	20	16	13	10	6
9	98	96	94	92	90	88	87	85	83	81	79	77	75	74	72	70	69	67	66	64	62	61	60	58	57	53	48	44	40	36	33	29	26	22	18	16	13	9
10	98	96	94	92	91	89	87	85	83	82	80	78	76	75	73	71	70	68	66	65	63	61	60	58	57	53	49	45	42	38	34	31	28	24	21	18	15	12

	0.5	1.0	1.5	2.0	2.5	3.0	3.5	4.0	4.5	5.0	5.5	6.0	6.5	7.0	7.5	8.0	8.5	9.0	9.5	10.0	10.5	11.0	11.5	12.0	12.5	13.0	13.5	14.0	14.5	15.0	15.5	16.0	16.5	17.0	17.5
12	95	91	87	82	78	74	70	66	62	59	55	51	48	44	41	38	35	31	28	25	22	19	16	13	11	8	5	3							
14	96	91	87	83	79	75	72	68	64	61	57	54	50	47	44	41	38	35	32	29	26	23	20	18	15	12	10	7	5	3					
16	96	91	88	84	80	76	73	69	66	62	59	56	53	49	46	43	40	37	35	32	29	27	24	21	19	16	14	12	9	7	5	3			
18	96	92	88	85	81	77	74	70	67	64	61	57	54	51	48	45	42	40	37	35	33	31	28	26	23	21	19	16	14	12	10	8	6	4	
20	96	92	88	85	81	78	75	71	68	65	62	59	56	53	50	48	45	42	40	38	35	33	31	29	27	25	23	20	18	16	14	12	10	8	6
22	96	93	89	86	82	79	76	72	69	66	63	60	57	54	51	49	46	44	42	40	37	35	33	31	29	27	25	22	20	18	16	14	12	10	8
24	96	93	89	86	83	80	77	73	70	67	65	62	59	56	53	50	47	45	43	41	39	37	35	32	30	28	26	24	22	20	18	16	14	12	10
26	97	93	90	87	84	81	78	74	71	69	66	63	60	58	55	52	50	48	46	43	41	39	37	35	33	31	29	27	25	23	21	19	17	15	13
28	97	93	90	87	84	81	78	75	72	69	67	64	62	59	57	54	52	49	47	45	42	40	38	36	34	32	30	28	26	24	22	20	18	16	14
30	97	94	90	87	84	82	79	76	73	70	68	65	63	60	58	55	53	51	49	47	45	43	41	39	37	35	33	31	29	27	25	23	21	19	17
32	97	94	91	88	85	82	79	77	74	71	69	66	64	61	59	57	54	52	50	48	46	44	42	40	38	36	34	32	30	28	26	24	22	20	18
34	97	94	91	88	85	83	80	77	74	72	70	67	65	62	60	58	55	53	51	48	46	44	42	40	38	36	34	32	30	28	26	24	22	20	18
36	97	94	91	88	86	83	80	77	75	72	70	68	65	63	61	59	57	54	52	50	48	46	44	42	40	38	36	34	32	30	28	26	24	22	20

Table 7-8. Psychrometric table for an atmospheric pressure of 350 torr or 467 millibars, equivalent to an altitude of about 6090 metres.

	0.2	0.4	0.6	0.8	1.0	1.2	1.4	1.6	1.8	2.0	2.2	2.4	2.6	2.8	3.0	3.2	3.4	3.6	3.8	4.0	4.2	4.4	4.6	4.8	5.0	5.5	6.0	6.5	7.0	7.5	8.0	8.5	9.0	9.5	10.0	10.5	11.0	11.5
-5	97	94	91	88	85	83	80	77	74	71	69	66	63	60	58	55	52	50	47	45	42	40	37	34	32	25	20	14	8	2								
-4	97	94	92	89	86	83	81	78	75	73	70	67	65	62	59	57	54	52	49	47	44	42	40	37	35	29	23	17	12	6								
-3	97	95	92	89	87	84	81	79	76	74	71	69	66	64	61	59	56	54	51	49	47	44	42	40	37	32	26	20	15	10	5							
-2	97	95	92	90	87	85	82	79	77	75	72	70	67	65	63	60	58	55	53	51	49	46	44	42	40	34	29	24	18	13	8	3						
-1	97	95	92	90	87	85	83	80	78	75	73	71	68	66	63	62	59	57	55	53	51	48	46	44	42	37	32	26	21	17	12	7	3					
0	98	95	93	90	88	86	83	81	79	76	74	72	70	67	65	63	61	59	56	54	52	50	48	46	44	39	34	29	24	20	15	10	6	2				
1	98	95	93	91	88	86	83	81	79	77	74	72	69	68	65	64	62	59	58	56	54	52	50	48	46	41	36	32	27	22	18	14	9	5	1			
2	98	95	93	91	89	86	84	82	80	78	75	73	70	69	67	65	63	61	58	57	55	53	52	50	48	43	38	34	29	25	21	17	12	8	4			
3	98	96	93	91	89	87	85	82	81	78	76	74	71	70	68	66	64	63	61	59	57	55	53	51	49	45	40	36	32	27	23	19	15	11	8	4		
4	98	96	94	91	89	88	85	83	81	79	76	74	71	71	69	67	66	64	62	60	58	56	55	53	51	47	42	38	34	30	26	22	18	14	11	7	3	
5	98	96	94	92	90	88	85	84	81	79	77	75	72	71	70	68	67	65	63	62	59	58	57	54	54	48	44	40	36	32	28	24	21	17	13	10	6	3
6	98	96	94	92	90	88	86	84	82	80	78	76	73	72	71	69	68	66	64	63	61	59	58	55	55	50	46	42	38	34	30	26	23	19	16	12	9	6
7	98	96	94	92	90	89	86	85	83	80	78	77	74	73	71	70	68	67	65	64	62	60	59	57	56	51	47	43	39	36	32	29	25	22	18	15	12	9
8	98	96	94	93	90	89	87	85	83	81	80	78	75	74	72	71	69	68	66	65	63	61	60	59	56	52	49	45	41	38	34	31	27	24	21	17	14	11
9	98	96	94	93	91	89	87	85	84	82	80	78	75	75	73	72	70	68	67	66	64	62	60	59	57	54	50	46	43	39	36	32	29	26	23	20	17	14

	0.5	1.0	1.5	2.0	2.5	3.0	3.5	4.0	4.5	5.0	5.5	6.0	6.5	7.0	7.5	8.0	8.5	9.0	9.5	10.0	10.5	11.0	11.5	12.0	12.5	13.0	13.5	14.0	14.5	15.0	15.5	16.0	16.5	17.0	17.5
10	95	91	87	82	78	74	70	66	62	58	55	51	48	44	41	37	34	31	28	25	22	19	16	13	10	7	5	2							
12	96	91	87	83	79	75	71	68	64	60	57	53	50	47	43	40	37	34	31	28	25	23	20	17	15	12	9	7	5	2					
14	96	92	88	84	80	76	73	69	65	62	59	55	52	49	45	42	39	37	34	31	29	26	24	21	18	16	14	11	9	7	4	2			
16	96	92	88	84	81	77	74	70	67	64	60	57	54	51	48	45	42	39	37	34	32	29	27	24	22	20	17	15	13	11	9	6	4	2	
18	96	92	89	85	81	78	74	71	68	65	62	58	55	52	49	46	44	41	38	36	34	31	29	26	24	21	19	17	15	12	10	8	6	5	
20	96	93	89	85	82	79	75	72	69	66	63	60	56	53	51	48	45	42	40	37	35	32	30	28	26	23	21	19	17	15	13	11	10	8	6
22	96	93	89	86	82	79	76	73	70	67	63	60	58	55	52	49	47	44	42	39	37	34	32	30	28	26	23	22	20	18	16	14	13	11	10
24	97	93	90	86	83	80	77	74	71	68	65	62	59	56	54	51	49	46	44	41	39	37	35	32	30	28	26	24	22	20	18	17	15	14	13
26	97	93	90	87	83	81	77	74	72	69	66	63	61	58	55	53	50	48	45	43	41	38	36	34	32	30	28	26	24	23	21	19	18	16	14
28	97	94	90	87	84	81	78	75	72	70	67	64	62	59	56	54	51	49	47	44	42	40	38	35	33	31	29	28	26	24	23	21	19	19	17
30	97	94	91	88	84	82	78	76	73	70	68	65	63	60	57	55	52	50	48	46	43	41	39	37	34	33	31	29	28	26	24	23	21	21	19
32	97	94	91	88	85	82	79	77	74	71	69	66	64	61	58	56	54	51	49	47	45	43	40	38	36	35	33	32	29	28	26	25	23	23	21
34	97	94	91	88	85	83	80	77	75	72	70	67	65	62	59	57	54	52	50	48	46	44	41	39	38	36	35	32	31	29	28	26	25	25	23
36	97	94	92	88	86	83	80	78	75	73	70	68	66	63	61	59	55	54	51	49	47	45	42	40	39	37	36	34	32	31	29	28	27	27	27

Table 7-9. Psychrometric table for an atmospheric pressure of 300 torr or 400 millibars, equivalent to an altitude of about 7150 metres.

	0.2	0.4	0.6	0.8	1.0	1.2	1.4	1.6	1.8	2.0	2.2	2.4	2.6	2.8	3.0	3.2	3.4	3.6	3.8	4.0	4.2	4.4	4.6	4.8	5.0	5.5	6.0	6.5	7.0	7.5	8.0	8.5	9.0	9.5	10.0	10.5	11.0	11.5	12.0
-5	97	95	92	89	86	84	81	79	76	73	71	68	66	63	61	58	56	54	51	49	46	44	42	39	37	31	26	20	15	10	4								
-4	97	95	92	90	87	84	82	79	77	74	72	70	67	65	62	60	58	55	53	51	48	46	44	42	39	34	29	23	18	13	8	3							
-3	97	95	92	90	87	85	83	80	78	75	73	71	68	66	64	61	59	57	55	53	50	48	46	44	42	36	31	26	21	16	12	8	3						
-2	98	95	93	90	88	85	83	81	78	76	74	72	69	67	65	63	61	58	56	54	52	50	48	46	44	38	34	29	24	19	15	10	6	1					
-1	98	95	93	91	88	86	84	81	79	77	75	73	70	68	66	63	61	59	56	56	54	52	50	48	46	41	36	31	27	22	18	13	9	5					
0	98	95	93	91	89	86	84	82	80	77	76	73	71	70	67	65	63	61	59	57	54	53	51	49	48	43	38	34	29	25	21	16	12	8	4				
1	98	96	93	91	89	86	85	82	80	78	76	74	72	71	68	66	64	62	61	59	55	55	53	51	49	45	40	36	32	27	23	19	15	11	8	3			
2	98	96	93	91	89	87	85	83	81	79	77	74	73	72	69	67	65	64	62	60	57	56	54	53	51	46	42	38	34	30	26	22	18	14	10	7	3		
3	98	96	94	92	90	87	86	83	82	79	78	75	73	73	70	68	66	65	63	61	58	57	56	54	52	48	44	40	36	32	28	24	20	17	13	10	6	3	
4	98	96	94	92	90	88	86	84	82	80	78	76	74	74	71	69	67	66	64	63	59	58	57	55	54	50	45	42	38	34	30	26	23	19	16	12	9	6	3
5	98	96	94	92	90	88	86	85	83	80	79	77	75	74	72	70	67	67	65	63	60	59	58	57	55	51	47	43	39	36	32	29	25	22	18	15	12	9	6
6	98	96	94	92	90	89	87	85	83	81	79	77	76	75	73	71	69	67	66	64	61	60	59	58	56	52	48	45	41	37	34	30	27	24	21	17	14	11	8
7	98	96	94	93	91	89	87	85	84	82	80	78	77	75	73	72	70	68	67	65	63	62	60	59	57	53	50	46	43	39	36	32	29	26	23	20	17	14	11

	0.5	1.0	1.5	2.0	2.5	3.0	3.5	4.0	4.5	5.0	5.5	6.0	6.5	7.0	7.5	8.0	8.5	9.0	9.5	10.0	10.5	11.0	11.5	12.0	12.5	13.0	13.5	14.0	14.5	15.0	15.5	16.0	16.5	17.0
8	95	91	87	82	78	74	70	66	62	58	55	51	47	44	41	37	34	31	28	25	22	19	16	13	10	8	5	2						
10	96	91	87	83	79	75	71	67	64	60	57	53	50	47	43	40	37	34	31	28	25	23	20	17	15	12	10	9	5	2				
12	96	92	88	84	80	76	72	69	65	62	59	55	52	49	46	43	40	37	34	31	29	26	24	21	18	16	14	11	9	5	2			
14	96	92	88	84	81	77	74	70	67	63	60	57	54	51	48	45	42	39	37	34	32	29	27	24	22	20	17	15	13	11	7	5	2	
16	96	92	89	85	81	78	75	71	68	65	62	59	56	54	51	48	45	42	40	37	34	32	30	27	25	22	20	18	16	14	11	9	7	3
18	96	92	89	85	82	79	76	72	69	66	63	60	57	54	53	50	46	44	41	39	36	34	32	30	28	26	24	21	19	17	14	12	10	7
20	96	93	89	86	83	79	76	73	70	67	64	61	59	56	54	51	48	46	43	41	39	36	34	32	30	28	26	24	22	20	18	15	13	10
22	96	93	90	86	83	80	77	74	71	68	65	63	60	57	55	52	51	48	45	43	41	38	36	34	32	30	28	26	25	21	19	16	13	
24	97	93	90	87	84	81	78	75	72	69	66	64	61	59	56	54	51	49	47	45	42	40	38	36	34	32	30	27	25	23	22	20	18	

Hair hygrometers are the most venerable and inexpensive instruments for the measurement of humidity. Since the extension of a hair is directly proportional to the logarithm of the relative humidity above 10% (6), direct-reading instruments are easily constructed. Their disadvantages are manifold and well investigated, but it appears that they can be used without correction for altitude.

Other instruments based on similar materials may also give results equal to the psychrometer. One German instrument, for example, claims a precision of '2% plus or minus' at all temperatures from 0° to 110° (2). It seems probable that such an instrument would operate nearly independently of atmospheric pressure and temperature. However, the manufacturer claims no knowledge of its performance in highland areas.

It is interesting to note that the humidity monitors in most manned spacecraft must function at reduced pressures. NASA used dew point monitors for measuring water content of the air in both the Apollo and the Skylab missions. Since all humidity measurements are interconvertible, direct measurements of dewpoints can be converted to relative humidity. Although non-automated measurements of dew point are relatively precise, they can be made by a person isolated from all other equipment.

The enterprising scientist can construct a gravimetric apparatus for high precision measurements. By the use of weighed dessicant tubes he can absorb and weigh the moisture in ambient air and in saturated air. Spencer-Gregory and Rourke call this a 'chemical psychrometer', and present a design for construction in chemical laboratories (6).

2. References

1. Deutschen Wetterdienst. 1964. Aspirations-Psychrometer-Tafeln. 4th enlarged ed. Braunschweig.
2. Giddings Jr. L. E. 1972. La quimica en la altura. III. La medicion de humedad. *Afinidad*, 24: 869–874; IV. Tablas psichrometricos para la altura. *ibid.*, 31: 935–945.
3. Marvin, D. E. 1941. Psychrometric tables for obtaining the vapour pressure, relative humidity and temperature of the dew point from readings of the wet and dry-bulb thermometer. US Supt. Documents. Washington DC.
4. Rechard, P. A. 1967. Psychrometric tables for Wyoming (High elevations). Water Resources Series No. 6. US Dept Int. Washington DC.
5. Psychrometric Tables. Canada Dept. Transport, Met. Branch Toronto. Books 1 and 3.
6. Spencer-Gregory, H. & E. Rourke. 1957. Hygrometry. London: Lockwood. *vide* p. 184.

I For the engineer

L. E. Giddings

The materials presented in other chapters of this book will be helpful to many kinds of engineers. Still, there are a number of problems of great interest to engineers which are not of consequence to others. A few of these are discussed in this chapter.

1. Azeotropes

One of the most vexing of problems concerns the behaviour of azeotropes. For the average well-trained chemical engineer, there is no easy way of understanding the composition and boiling point changes of azeotropes as the pressure changes. Most tabulated data specify sea level atmospheric pressure; but since azeotropes reflect non-idealities of solutions, it is not obvious how they change to meet the lower pressures of the highlands.

As a practical example, water and ethanol form an azeotrope at 95% ethanol at sea level. At 500 torr, corresponding to an altitude somewhat over 3300 m, it increases to 99%. For most of us, the reason for this increase is not immediately apparent.

One first needs access to azeotropic data. By far, Horsley's (8) compendium and only supplement (9) are the most complete sources of data (Supplements need to be issued to cover the chemical literature since about 1960, but none seems to be currently planned.) Both of these fundamental references are long out of print. One may search the standard chemical handbooks, but these are very deficient. Literature surveys through Chemical Abstracts (3) will locate all published azeotropic data, but this is quite a lengthy procedure.

Horsley's (8, 9) books show some practical ways of dealing with reduced pressures of azeotropes. He demonstrates, for example, that the vapour pressure curves are linear when plotted on a Cox scale (4).

2. Combustion

Another problem is the behaviour of flames and combustion in general in high areas. Of course, calculations of stoichiometry of flames is simple enough, but many subtle problems occur. As an example, the author vividly remembers an alcohol burner that would not support a steady flame in his laboratory at 3040 m; presumably it would have functioned well at sea level since it was furnished by a reliable dealer.

Since the percentage composition of air does not vary with terrestrial altitudes, its oxygen content varies only with its pressure (and, to a minor extent, with its humidity). As a result, an engineer may use the tables and equations of Chapter II to correct air-fuel ratios to highland conditions.

However, most practical problems are more subtle. The most we can do in this book is set a descriptive theoretical background. Understanding the various kinds of flames, and how a few of their functions are affected by pressure, an engineer should be better equipped to cope with the problem of combustion at high terrestrial altitudes.

As a start, we must recognize that practical flames can be classified as predominantly premixed or predominantly diffusion flames. Premixed flames are well represented by the Bunsen flame, in which premixed gases flow up a burner tube at a rate which exceeds the burning velocity of the mixture, a steady flame being maintained above the burner top. Premixed flames are used for domestic heating, for gas flames, for cooking, etc. They have been studied more than diffusion flames because they are simpler, and because they can yield information about the gas mixture, such as its burning velocity and temperature.

Predominantly diffusion flames depend on interdiffusion of air and fuel. Examples include simple gas jets in air, and flames on wicks. In large diffusion flames, mixing may be due to turbulence and other factors as well as diffusion. Heat transfer plays a role in the production of flammable vapour in some of these systems, as, for example, in a candle.

Diffusion flames are more difficult to study than premixed flames, but the following are some effects of pressure. As the pressure decreases, the rate of diffusion of gases increases, but the rate of reaction decreases. Therefore, there is more mixing, giving more premixed condition to the diffusion flames. By comparison, diluting burning gases with an inert gas decreases reaction rate without increasing diffusion; in the normal case, nitrogen of the air performs this role as diluent. Finally, for a given flame, such as that of a candle, the flame becomes longer as the atmospheric pressure is reduced.

Chemical considerations are more complex in the diffusion flame. Few physical constants can be measured, and there is not even a burning velocity in the ordinary sense. Although these flames are important industrially, their detailed study is beyond the scope of this chapter. The interested reader might start with references 5, 6 and 7 as an introduction; in particular, the fascinating book by Gaydon (5) treats of a very wide variety of flames in an interesting way.

In representative premixed flames, the premixed gases travel up the burner tube at a rate faster than the velocity of the flame. If the flow of gas is too slow, the flame will 'strike back' (travel back into the tube) or 'quench' (die). There is some additional stabilization of the flame at the rim of the burner, and there is a quenching effect inside the tubes at the wall.

If the gas flow is too rapid, the burning velocity will be exceeded and the flame will rise until it takes up a position above the burner rim where the burning velocity and the flow are just equal. A stable lifted flame is possible because the gas flow lines will diverge slowly, giving slower flow, while the burning velocity increases with distance from the rim. However, if the flow of gas is too high, the flame will blow off.

Quenching effects are important in determining flame stability. Pressure effects on quenching are well known, mostly studied in relation to the quenching diameter of a burner, the burner diameter at which a flame dies rather than flashing back. In general, the quenching diameter increases as the gas pressure is reduced, being roughly inversely proportional; this is also found to be true for stability regions of premixed flames. That is to say, in the highlands, burners might well have greater diameters.

Flame speed of premixed flames does not much vary with pressure. Since the pressure of limiting flames varies inversely with the burner diameter, the mass rate of burning tends to increase roughly inversely with pressure with flames at limiting conditions. This should have little significance for most flames within normal combustion limits of pressure and burner diameter; but its suggests that faster pumping of reactor gases would stabilize some flames that are brought to a lower limit of stability by a decrease in pressure.

The ignition of flames is also complicated and well studied. In general, the spontaneous ignition temperature increases with a decrease in pressure. For ignition by external means, the minimum energy varies about as $1/P^2$; that is, more energy is required at lower pressures. Ignition also depends strongly on other conditions, such as gas flow.

3. More Complex Flames

The expressions 'predominantly diffusion' and 'predominantly premixed' suggest that even simple flames are quite complex. Indeed, though the inner cone of a Bunsen flame may well be premixed, as most students of elementary chemistry are taught, the outer cone is definitely a diffusion flame, largely controlled by interdiffusion of combustible gases and air.

Certainly, other important flames are much more complex. An important form of industrial combustion involves sprays of droplets of combustible petroleum liquids; the reader should understand that treating such flames, both in their premixed and their diffusion aspects, is extremely complex. Consider also the so-called 'cool flames', which can be as cool as 200° or 300°. Gaydon gives the flavour of our understanding of these and many other flames (5).

Clearly, this discussion cannot cover more than the fringe of the problem of combustion at ambient conditions in the highlands. Flames are very complex, and the practical problems of combustion in the highlands have not been documented in the technical literature.

4. Engines

It is experience that internal combustion engines behave differently in the highlands. Automobile travellers find that their vehicles change in performance as they climb from sea level to highlands. These problems were studied in earlier days because of their impact on the operation of aircraft.

The simplest correction factor that is normally applied concerns engines that ignite fuel mixtures with sparks. If well adjusted to the altitude, they should contain the same relative composition of fuel and air as at sea level, but the mass of air in the highlands would be less. Therefore their maximum power is proportional to the weight of air (or mixture) inducted into the cylinder. The following correction factor will, therefore, correct the maximum power obtainable from the motor (6):

$$\text{factor} = \frac{\text{operating pressure}}{\text{standard pressure}} \times \sqrt{\frac{\text{standard absolute ambient temperature}}{\text{operating absolute ambient temperature}}}$$

For precise use, the partial pressure of water vapour must be subtracted from the total pressure. Obviously, since the pressure is the stronger of the two variables in this equation, spark-ignited engines would lose considerable power in the highlands.

Engines ignited by compression, such as diesel engines, maintain a constant flow of air into which fuel is injected. The ratio of air to fuel is not as nearly constant as with spark-ignited engines. For this reason, these engines often do not use all the air present in the cylinder, and therefore the above correction cannot be precise. However, a similar type of equation can be constructed for any given engine.

The obvious solution to the lower availability of air at high altitude is to pump air into the engine. Such an air pump, called a supercharger, allows the engine to obtain the same mass of air at high altitudes as it does at sea level. Properly operated, a supercharger should allow a motor to maintain its power at a constant level from sea level to an altitude determined by its design.

For engines using natural gas and similar combustibles, Campbell summarizes altitude corrections (2). He presents the following as typical lowering of ratings:

High Speed Engines:
>non-turbocharged – 3% for each 300 m (1000 ft)
>>above sea level
>turbocharged – 2% for each 300 m above sea level

Larger Integral Engines:
>no loss to 500 or 750 m (1500 to 2500 ft)
>>then 3% loss per 300 m for natural aspirated
>>engines
>or 2% loss per 300 m for turbocharged engines
>>(some large engines are designed for no correction
>>to 2000 m)

Campbell (2) also presents Fig. 8-1, derived from air density curves. He states that in the absence of specific data for a given engine, either this figure or the information given above can be used for planning purposes.

These data are often available for specific engines. As an example, Fig. 8-2 presents the curve furnished for use with specifications of the gas compressors and mechanical drive units of Solar Turbines International. The same company also furnishes specific instructions and examples for using it to obtain maximum power, fuel consumption and exhaust heat with its equipment.

Internal combustion motors are known to function best with gasoline of a certain octane rating. The optimum octane requirements also vary with altitude. The author could not find any general rules, but the manual for one automobile stated that its octane requirements decrease one octane unit for each increase of 600 m of altitude (1).

This short discussion bypasses all considerations of engine construction and adjustment that may help performance at altitude. Certainly carburettors are often made larger to allow less resistance to the flow of the air. Some adjustments can be made, such as advancing or retarding the spark.

5. Gravity

Gravity also varies with altitude. Though these variations are no doubt small, they are important to engineers of geologists, and perhaps to others. The discussion that follows is no more than an introduction based on the 'WMO Guide to Meteorological Instrument and Observing Practices' (11) and the book by Griffiths and King (7). The interested reader is urged to consult these and more specialized works.

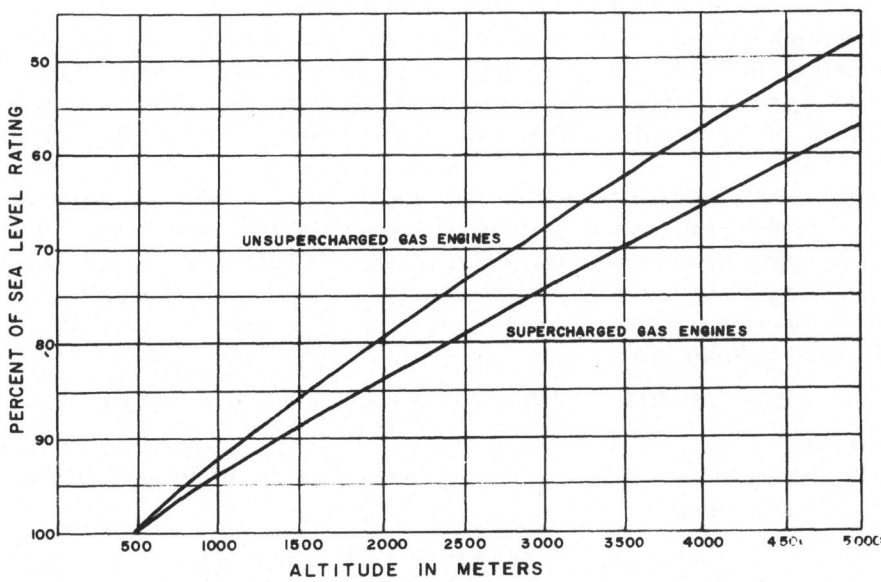

Fig. 8-1. Campbell's altitude corrections for gas engines. (Courtesy International Petroleum Institute, Ltd. and Natural Gas Producers Association.)

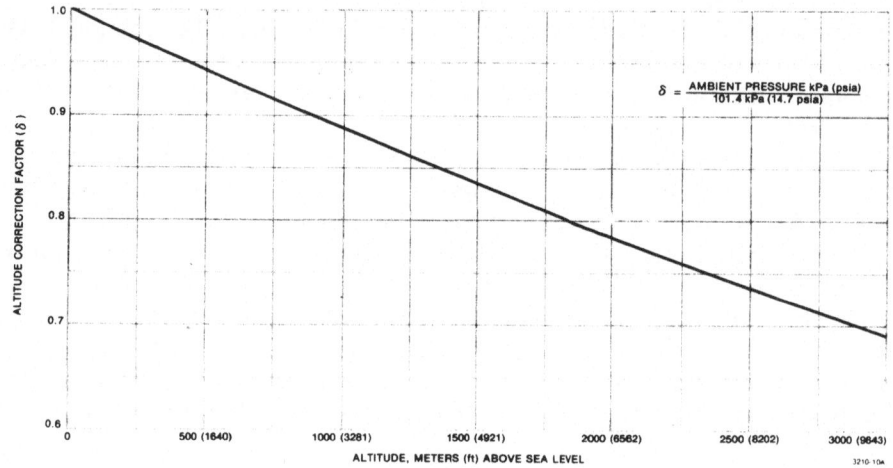

Fig. 8-2. Performance correction factor for altitude, for use with gas compressors and mechanical drive units. (Courtesy Solar Turbines, International.)

The acceleration due to gravity of the earth is about 980 cm sec^{-2}. If the earth were a perfect, homogeneous sphere, its value would be everywhere constant. However, since it is more nearly an oblate spheroid, generally taken to be a triaxial ellipsoid of with equatorial radius 21 km greater than the polar radius (this constitutes the so-called 'reference ellipsoid', which is about the same as the sea level surface of the earth).

Since the poles are nearer the centre of mass, gravity increases with an increase in latitude. At the same time it is decreased in the same direction by the rotation of the earth, which is a maximum at the equator, and which is at most about a third of the gravitational attraction.

For high altitude, two more factors must be considered, the free air correction and the intervening mass correction. The first compensates for height above sea level, because the observer is farther away from the centre of mass of the earth. The second corrects for the mass of rock that intervenes.

The following instructions from the WMO manual derive values based on the meteorological gravity system. The so-called Potsdam system, widely used by geodetic organizations, yields values 0.013 cm sec^{-2} greater that the meteorological system. The results, in any case, are approximate.

The theoretical value at mean sea level at geographical latitude ϕ is calculated as follows:

$$g_{\phi,0} = 980.616(1 - 0.0026373 \cos 2\phi + 0.0000059 \cos^2 \phi)$$

in cm sec^{-2}.

The local value at a given point on the surface of the ground at a land station is computed by:

$$g = g_{\phi,0} - 0.0003086H + 0.0001118(H - H')$$

where g = local calculated value at the given point, $g_{\phi,0}$ = theoretical value at mean sea level as calculated above, H = actual elevation of the given point in metres above mean sea level and H' = mean elevation in metres above mean sea-level of the actual surface of the terrain included within a circle whose radius is about 150 km, centred at the given point.

For greater precision one must use a gravimeter for comparisons with a base station of known gravity, or he can make calculations based on Bouguer anomalies. Details are beyond the scope of this book (7, 11).

6. References

1. Cadillac Motor Company, 1966. Cadillac Shop Manual, General Motors Corporation, Detroit (1965).
2. Campbell, John M. 1968. 'Gas Conditioning and Processing', International Petroleum Institute, Norman, Oklahoma.
3. Chemical Abstracts, American Chemical Society, Washington, D.C.
4. Cox, Edwin R. 1923. Pressure-Temperature Chart for Hydrocarbon Vapors, *Industrial and Engineering Chemistry*, 15: 592–3.
5. Gaydon, A. G. & H. G. Wolfhard. 1970. 'Flames', London: Chapman & Hall.
6. Glassman, Irving 1977. 'Combustion', New York: Academic Press.
7. Griffiths D. H. & R. F. King. 1965. 'Applied Geophysics for Engineers and Geologists', Chapter 6, London: Pergamon Press.
8. Horsley, L. H. 1952. 'Azeotropic Data, Tables of Azeotropes and nonazeotropes' Advances in Chemistry Series No. 6, American Chemical Society, Washington D.C.
9. Horsley, L. H. & W. Stamplin. 1962. Supplement to the above.
10. Walker, Jearl. 1978. The Amateur Scientist: The Physics and Chemistry Underlying the Infinite Charm of a Candle Flame, *Scientific American*, 238: 154–162.
11. World Meteorological Organization, 'Guide to Meteorological Instruments and observing Practices', 4th ed., sect. 3.8, Geneva (1971) WMO-No. 8 TP. 30.

The highlands in the space age

L. E. Giddings

It is common knowledge that the United States and the U.S.S.R. and perhaps others can survey any part of the Earth from space. Both countries launch satellites for 'classified purposes' as well as for more public purposes. Some details even of the military flights do reach the general public.

The same technology has proven useful to highland areas for peaceful purposes. This chapter will explore some ways of gathering information about highland areas from satellites and planes. It will serve as an introduction to the broad field of remote sensing, and show practical ways of relating it to the needs of higland areas. Particular attention will be paid to relatively simple and inexpensive ways of collecting useful information.

1. Remote Sensing of the Environment

The use of satellites and planes is especially pointful for gathering information about highland areas. High areas are generally not highly populated, and in many ways they tend to be remote and difficult of access on the surface. Often they are large, and in many areas, they are more free of clouds than nearby lowlands.

Early space photographs, such as the Gemini photographs of South America and the Apollo views of Tibet and Mexico (Plates 1 through 3) provided truly synoptic views of the highlands, and they are as informative as they were beautiful. For the first time, men could see the large portions of the upland plateaus as units, could see their relation to other geographical units, and could even appreciate the immensity of their resources. One photograph often provided more geographical information than was previously known about these areas.

The application of space technology to the study of the earth's surface has come to be known as 'remote sensing'. This has developed into a broad field for the study of the earth from remote 'platforms', the name given to the vehicle on which the instruments, or 'sensors', are mounted. References 13, 24, 29, 34, 35, 36 and 41 offer a useful introduction to this interesting new field.

Although most of the techniques developed for remote sensing are expensive or complex, many are not. Some are uniquely suited to the needs of high areas. In any case, much useful information is available to users at very low prices.

2. Instruments for Remote Sensing

Most studies of the earth from space are made with the electromagnetic spectrum, and the great majority are confined to visible and infrared

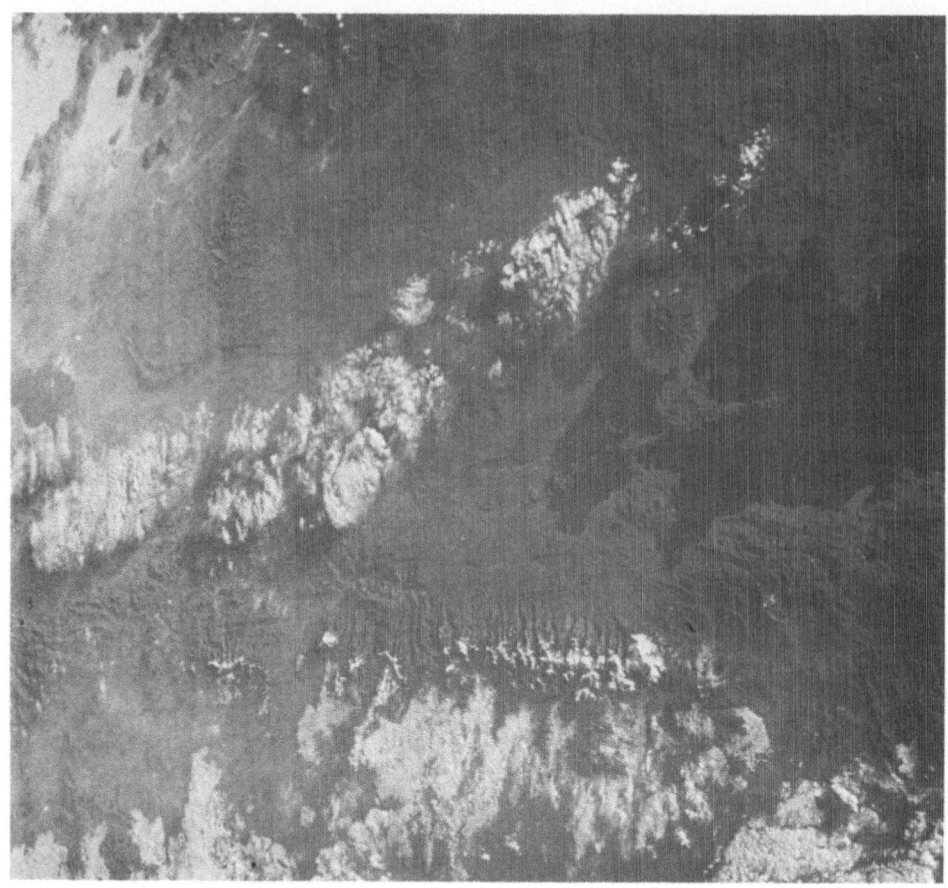

Plate 9-1. Gemini photograph of the Bolivian highlands (S65-45792).

radiation. Fig. 9-1 shows some remote sensing systems in use in the United States, mostly by the National Aeronautics and Space Administration of the United States, NASA. Acronyms are characteristic of operations of NASA, and those used in this figure are explained briefly in Table 9-1. The figure also traces the transmission properties of the atmosphere in broad outline, much as the figure in Chapter 3.

As the simplest, and certainly one of the most valuable and versatile forms of remote sensing, one may take photographs from a plane or manned satellite. Photography is limited to radiation in the ultraviolet and visible regions, plus a small portion of the near-infrared region. Since atmospheric aerosols cause strong scattering at short wavelengths, ultraviolet photographs are nearly useless, but visible and near-infrared photographs are quite valuable. Useful films for all these spectral ranges are quite easily found: colour, colour infrared, and panchromatic, orthochromatic, and infrared black and white films are all items of commerce, as are filters to isolate particular spectra regions.

102

Although photography is the most general system used for remote sensing, radiometry has some very distinct advantages. A radiometer, basically a device which measures radiant energy from a given area, is useful over a much wider spectral range than a camera. Indeed, most devices listed in Fig. 9-1 are radiometers, with the Return Beam Vidicon (a television camera) the chief exception.

The most useful radiometers for remote sensing are scanning devices; they measure energy at discrete points, or pixels, which can afterwards be assembled into the image of a scene. (Scanning radiometers might well be called electronic cameras.) Their data are usually recorded on computer tapes, which can afterwards be processed to produce photographic images or images on television screens. NASA's airborne Multispectral Scanner, for example, simultaneously records images in 24 spectra ranges, and the Multispectral Scanners of the Landsat satellites produce images in four spectral regions. Since these radiometers produce digital electronic data on computer tapes, computers can be used in their analysis.

Plate 9-2. Apollo photograph of the Tibetan highlands (AS7-1748).

103

Plate 9-3. The central plateau of Mexico (AS9-3685).

Remote sensing is not limited to devices which can produce images. The non-scanning spectrometers in Fig. 9-1 are used to measure properties of the atmosphere, incuding temperature, humidity and even composition. Still more interesting to highland areas are airborne gamma ray spectrometers, which can measure uranium, potassium and thorium content of the earths crust below (9); airborne magnetometers, which measure variations in the magnetic field of the earth, usually a function of iron content (10); and gravitometers, which measure variations in gravity, normally due to varia- tions in the geologic structure (11). In addition, sidelooking airborne radar, usually called SLAR, can be used to photograph large areas of the earth through clouds (4, 21, 42).

3. Platforms for Remote Sensing

Sensors are normally mounted on aircraft or on satellites. In practice, the field of view and the resolution of objects largely depend on the height of

104

REMOTE SENSOR WAVELENGTH COMPARISON

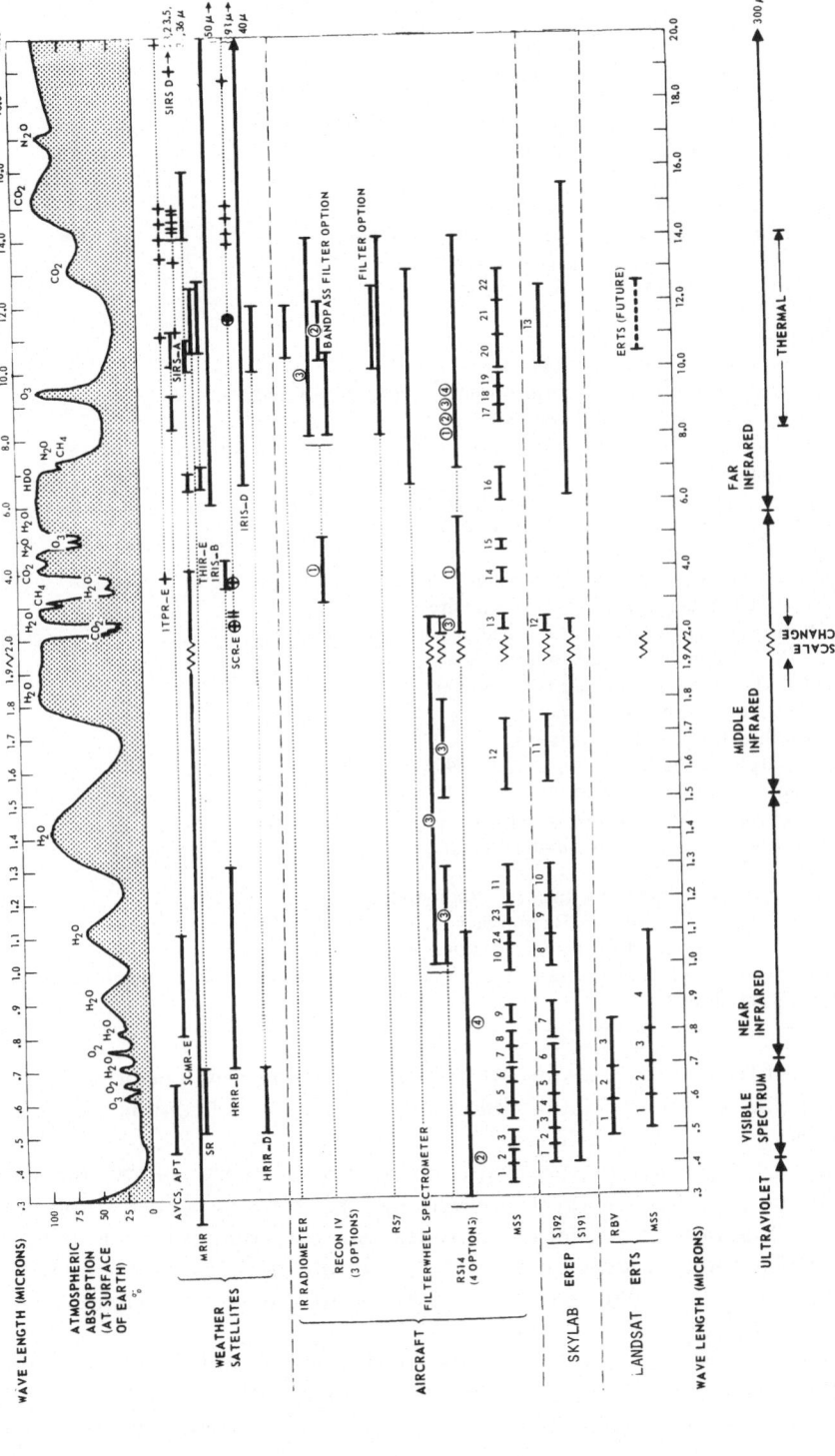

Fig. 9-1. Remote sensing systems.

PREPARED BY W.R. JOHNSON/PIMO/LEC 2/6.73

105

Table 9-1. Acronyms and special terms for remote sensing.

Apollo – an advanced series of manned satellites of NASA
APT – Automatic Picture Transmission
ATS-1, etc. – Applications Technology Satellites
EREP – Earth Resources Experiments of Skylab
EROS – Earth Resources Observations Systems, a data center
ERTS-1–the first Earth Resources Technology Satellite,
            renamed Landsat-1
Gemini – an early series of manned NASA satellites
GMS – Japanese geosynchronous meteorological satellite
GOES – Geostationary Orbiting Earth Satellite
HRIR – High Resolution Infrared Radiometer
IR – Infrared
IRIS – Infrared Interferometer Spectrometer
ITPR – Infrared Temperature Profile Radiometer
Landsat – the current series of earth resources technology satellites
Meteor – geosynchronous meteorological satellite of the U.S.S.R.
Meteosat – geosynchronous meteorological satellite of the European' Space Agency
MSS – Multispectral Scanner
NASA – the National Aeronautics and Space Administration
Nimbus-5, etc. – polar orbiting satellites, used as test beds
                for prototype sensors
NOAA – National Oceanic and Atmospheric Administration, also their
            polar-orbiting meteorological satellites
RBV – Return Beam Vidicon (in essence, a television camera)
RECON IV – scanning infrared radiometers
RS7 and RS14 – scanning infrared radiometers
SCR – Selective Chopper Radiometer
SIRS – Satellite Infrared Spectrometer
Skylab – a large manned satellite, used in 1973 and 1974
SLAR – Side Looking Airborne Radar
SMS – Synchronous Meteorological Satellite (similar to GOES)
Shuttle – a flexible reusable manned satellite
SR – Scanning Radiometer
THIR – Temperature Humidity Infrared Radiometer
VHRR – Very High Resolution Radiometer
VTPR – Vertical Temperature Profile Radiometer
WEFAX – Weather Facsimile, a relay system for transmitting images and information with the
SMS/GOES satellites

the platform. The state of the art of remote sensing now permits the use of any altitude to about 36,000 kilometres, the geosynchronous altitude for satellites.

The least expensive platforms are small aircraft. Every country has planes that can be used for low level photography, and most also have some planes that can function at moderate altitudes. Since reconnaisance is a normal military function, it only remains to substitute multispectral cameras for existing military reconnaisance cameras, or to use reconnaisance cameras with appropriate films and filters.

Because of costs, satellite systems are developed and flown only by the wealthier countries. However, within a few years, with the advent of routine flights in NASA's Space Shuttle, costs will be low enough for smaller countries to fly their own experiments. In any event, there will continue to be a wealth of satellite data made available to the international public by NASA at very reasonable prices. Details are given below, in the section on the EROS Data Center.

4. Multispectral Photography and Radiometry

The potential of photography and radiometry is not exploited by simple black and white photographs or images similar to them. Personal experience shows that ordinary colour films, with three primary colours, yield much more information. Logic suggests that more than three colours, or a different selection of colours, will somehow yield even more information about any given scene.

There has been a wealth of experimentation with various numbers and combinations of spectral channels. NASA's airborne multispectral scanner, a distinctly experimental device with 24 separate spectral channels, was used to test the usefulness of individual channels, as well as the potential of various combinations of them. They showed that our ability to distinguish and measure ground features tends to increase with the number of channels; it also increases with proper selection of channels for specific uses. However, the cost increases rapidly when we use more than three channels (the number implicit in colour or infrared colour photography). The use of more than three channels requires computer techniques instead of visual techniques.

Plates 9-3 through 9-6 show a single image in the four channels produced by the multispectral scanner of Landsat 1. These four images can be compared to black and white photographs taken with filters admitting only green, red and two bands of near-infrared radiation (although photographic films are not sensitive to one of the near-infrared bands).

The two near-infrared images are especially valuable for discriminating vegetation and water. Vegetation reflects near-infrared radiation much more efficiently than visible light, and water absorbs it better. Roads are often found to be more visible in the green image, and the red image often provides the best overall contrast for different features. It is apparent that each of the four Landsat images is useful, but for many uses, some kind of composite image is often convenient.

5. Integrating Several Spectral Bands

Ordinary colour photographs are composites of images in three spectral bands, or colours. The bands are chosen to reproduce the colours perceived by the human eye-brain system. However, blue light is not very useful from aircraft or satellite altitudes because of scattering, and near-infrared has proven to be very useful. For this reason, taking into account the technical limitations of the photographic process, a very useful colour infrared film

107

Plate 9-4. Landsat image in green (NASA 1306-16231-4).

has been developed in which near-infrared radiation is reproduced as red, red is reproduced as green, and green is reproduced as blue (this film was first developed for the detection of military camouflage from the air). In this way, an impressive amount of information is presented in a way that can easily be interpreted by the human eye-brain system.

Plates 9-8 and 9-9 were reconstituted from Landsat-1 imagery using this 'false colour infrared film' code for colours. With practice, this type of imagery is easily interpreted. Healthy vegetation is always red, but diseased vegetation will normally have some other colour; water is normally blue or black; sand is tan; and various other objects, such as roads and urban areas, often have characteristic colours.

In preparing images such as Plate 9-8 and 9-9, any spectral band can be assigned to any of the three primary colours, but assigning more than three is difficult. Arbitrary combinations of bands can be assigned to each of the three primary colours, but the resulting images tend to be hard to interpret.

108

Plate 9-5. Image in red light (NASA 1306-16231-5).

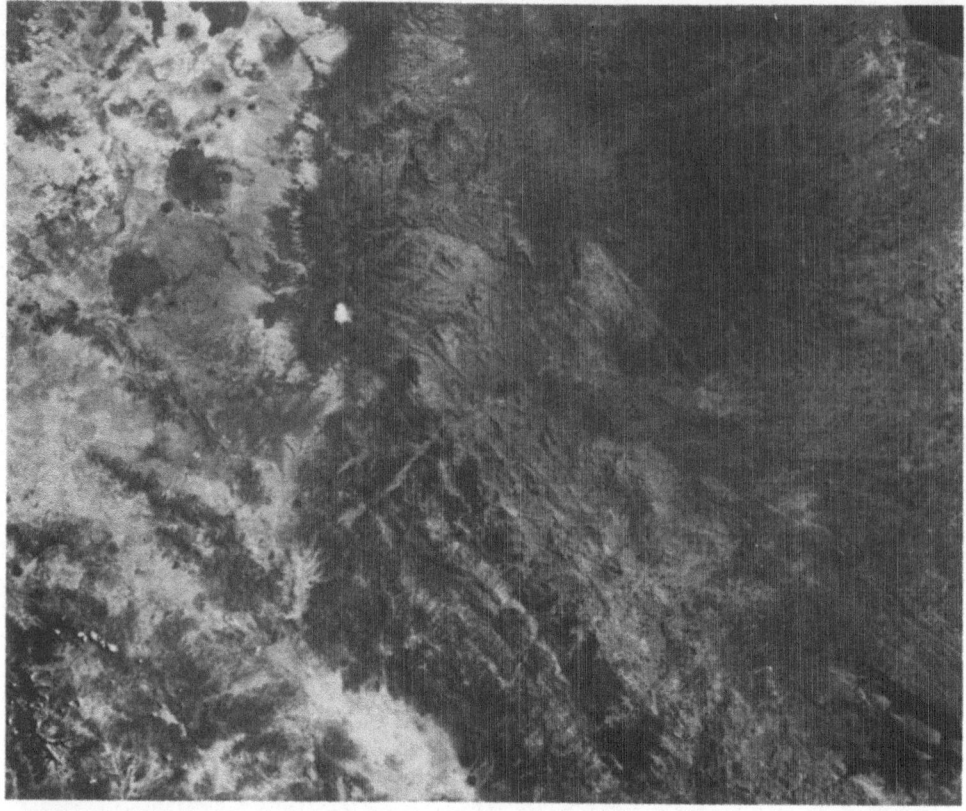

Plate 9-6. Image in near infrared radiation, 0.7 to 0.8 μm (NASA 1306-16231-6).

However, image-analysis computer systems are able to deal with many spectral bands at a time. They allow an analyst to use as many bands as prove necessary to distinguish features in a scene.

As a practical tool, if resources are limited, photographic images of Landsat scenes from the EROS Data Center will probably be the most useful for most projects using remotely sensed data. However, a person should not overlook the possibilities of ordinary colour or colour infrared film from hand-held cameras in light planes.

6. Landsat False Colour Infrared Image

This view of Ethiopia contains several high areas as well as the African Rift Valley. It was prepared from single channel data similar to plates 9-4, 9-5 and 9-7. Note that the highest areas are tinged with red, indicating that they have more vegetation than low areas. These Rift lakes are quite diverse, varying from fresh to saline and clear to turbid. Centre point coordinates are given on the second line of the legend, next to the date. This is a relatively inexpensive way of studying small areas of highlands. Colour infrared film can be processed by many laboratories accustomed to handling conventional

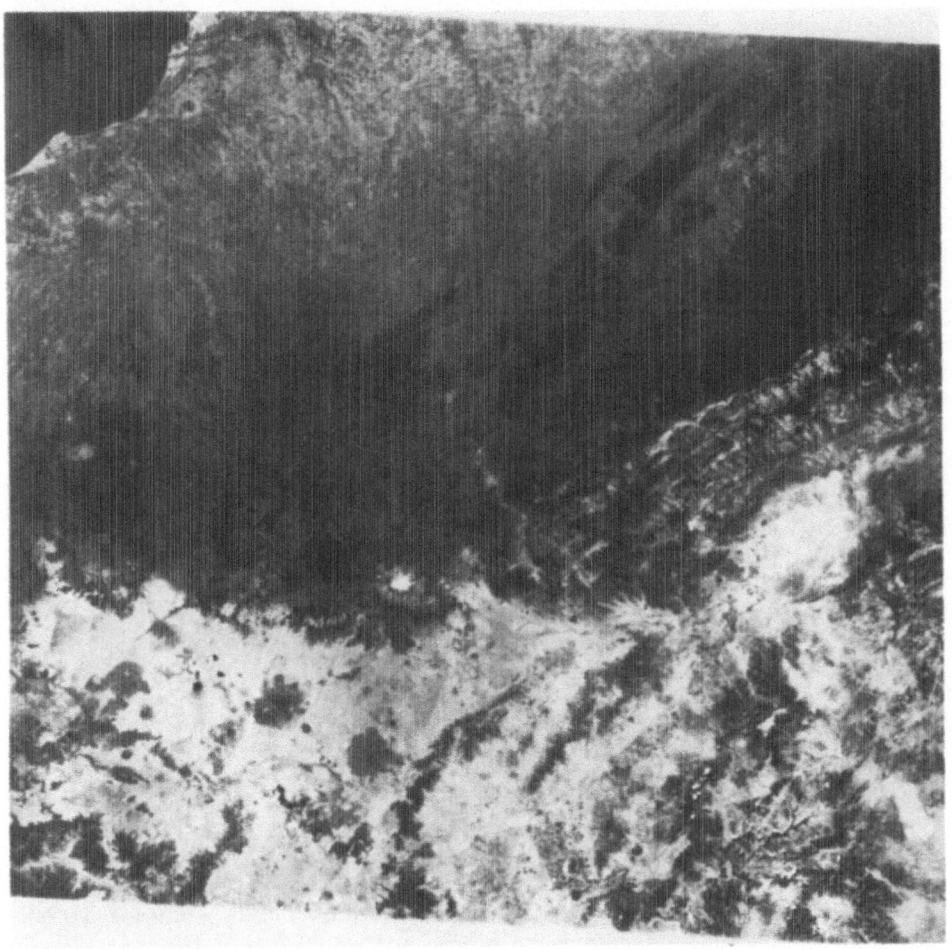

Plate 9-7. Image in near infrared radiation, 0.8 to 1.1 μm (NASA 1306-16231-7).

colour films, and hand-held cameras with filters can be handled well with a minimum of training.

Scanning radiometers also promise to become a relatively inexpensive tool. At the time of writing this, radiometers similar to those in Landsat are being offered at prices that are no longer astronomical. However, one still needs access to computers for processing raw data from multispectral scanners.

7. Photographs from Space

Most highland areas can be seen in the photographs taken by astronauts of the several manned missions. Most were taken as part of pre-planned experiments, but some were taken at the initiative of the astronauts. Since some were taken as early as 1965, they constitute a historical base that might be useful in some studies.

Most of the Gemini, Apollo and Apollo–Soyuz photographs were taken with hand-held cameras. All Apollo 6 photographs and the multispectral

111

Plate 9-8. Landsat false colour infrared image prepared from Plates 9-4, 9-5 and 9-6.

photographs from Apollo 9 and Skylab were taken with cameras mounted on the spacecraft. Most were taken with fairly standard Ektachrome films, but the Apollo 9 multispectral photographs were taken with colour infrared film and three combinations of black and white films with filters. Skylab multispectral photographs were taken in six exposures for each scene: visible colour, false colour infrared and four combinations of black and white films and filters.

In the near future, photographs should become available from NASA's Shuttle. Indeed, NASA will probably respond to requests for photographs needed for studies of highland areas. However, at this writing, the cameras to be carried on Shuttle have not yet been identified.

There are many references on photography that will be helpful in studies of earth resources (15–16).

8. The Landsat Satellites

In July of 1972, the first Earth Resources Technology Satellite, then called ERTS-1 but since renamed Landsat-1, was successfully launched into a precise

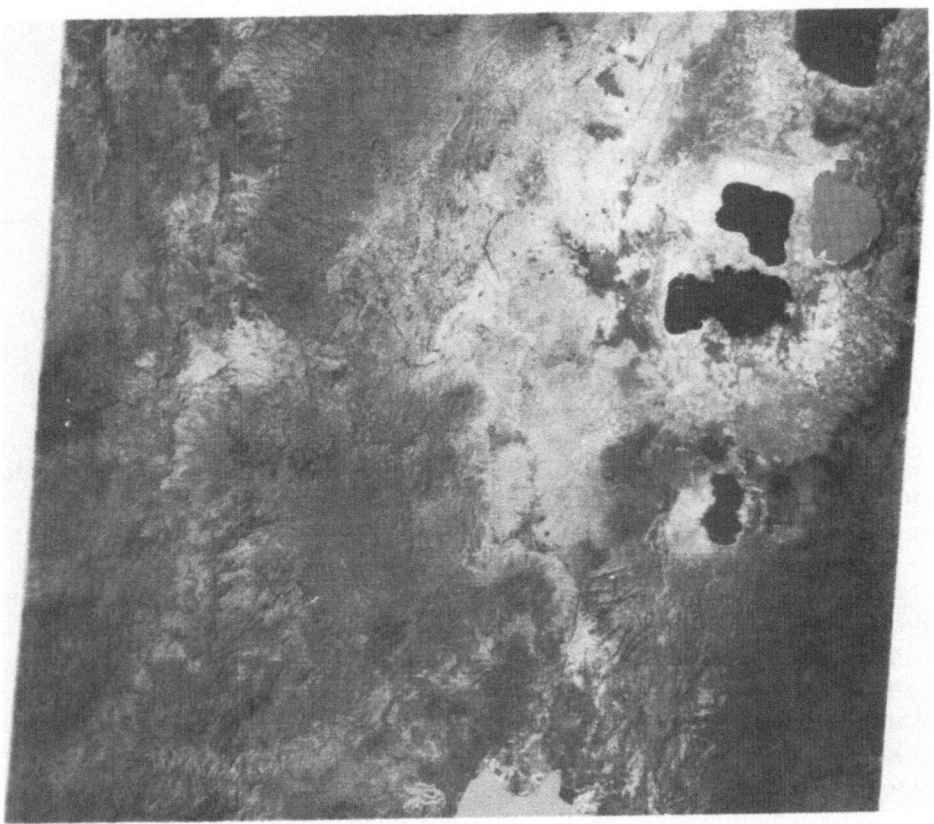

Plate 9-9. Landsat false colour infrared image.

polar orbit. In retrospect, this probably marked the beginning of the wide use of satellite data for studying the surface of the earth. Multispectral Landsat-1 images were taken liberally, and they were made easily available, at nominal cost, to all who wanted them. They proved extremely useful in a wide variety of applications, and their usefulness is being extended constantly.

Landsat-2 was launched in 1975 as some systems failed on Landsat-1. It also provided very fine multispectral images, so useful that a third satellite, Landsat-3, was launced in 1978.

Landsat images are taken whenever the satellite passes over the United States, but they are taken elsewhere only for approved projects. It is normally necessary to arrange with the NASA Landsat Project Office to ensure that images will be taken.

Once they are taken, images are always available for purchase through the EROS Data Center, which will be mentioned in greater detail below, or with certain other organizations charged with regional responsibilities. The archives of EROS probably contain images of every highland area of the earth; they certainly contain useful images of a large fraction of the land surface of

113

the earth, with exception of certain areas that are normally obscured by clouds.

9. Polar Orbiting Meteorological Satellites

Although photographs and Landsat images of the earth are spectacular because of their resolution of detail and the spectral coverage, they are probably equalled in importance by the less spectacular images from meteorological satellites. Polar-orbiting meteorological satellites take complete images of the earth twice daily, and geosynchronous satellites furnish images every half hour. The products of meteorological satellites are also freely available to the public, mostly at nominal cost.

NOAA/ITOS/TIROS and Nimbus satellites pass over both poles of the earth, in sight of all areas by day and by night. All produce horizon-to-horizon images of the earth in thermal-infrared radiation on both passages, and visible images (equivalent to a panchromatic black and white photograph) by day. Composite images of the whole earth are prepared daily for NOAA satellites and published monthly by the U.S. National Oceanic and Atmospheric Administration (12).

Plates 9-10 and 9-11 present images from a NOAA satellite. They are raw images, uncorrected for the foreshortening of areas as seen from the satellite. Note that the contrast between water and land is vivid in the visible image (Plate 9-9), but it is very low in infrared radiation if the temperatures on land are not much different from the temperatures of the water. Vegetation registers dark in visible light because of the reflectivity of vegetation. Many features of geography are especially evident in thermal infrared radiation because of temperature variations; valleys in the high plateau of Mexico can be easily discerned in Plate 9-11 because they are warmer than the plateau, but they cannot be seen in the visible image (Plate 9–10). In addition, although we are accustomed to white clouds and dark shadows, the shadows tend to be light in colour in Plate 9-11 since they are colder.

NOAA satellites also have a Vertical Temperature Profile Radiometer, VTPR (30, 38). This interesting device, essentially a specialized multichannel infrared spectrometer, gives raw readings that can be converted to a temperature profile of the atmosphere from ground level to the upper troposphere. These data are now available by ordinary Automatic Picture Transmission to any user, along with the normal imagery data. Conversion of the raw data to an actual temperature profile, a very complex computational task requiring large computers, is described in McMillan et al. (30).

NIMBUS satellites are research devices used as test beds for experimental sensors. A full account of sensors included on recent NIMBUS satellites is beyond the scope of this book (39). However, it seems appropriate to mention that some of the recently flown sensors prepare images from radiation of other portions of the electromagnetic spectrum. For example, the Temperature-Humidity Infrared Radiometer, THIR, gathers images at

114

Plate 9-10. Visible image from the very high resolution radiometer of the NOAA-4 satellite.

6.5 to 7.0 μm as well as in the thermal infrared region from 10.5 to 12.5 μm. The former image is very sensitive to atmospheric humidity and allows a calculation of water vapour in the lower troposphere. Other imaging sensors, such as the Electronically Scanning Microwave Radiometer, ESMR, constitute first attempts to determine the usefulness of the microwave region of the spectrum; ESMR promises to be useful for monitoring liquid water, such as rain, surface water or soil moisture (3, 37).

NIMBUS satellites include many non-imaging sensors. Developmental models of spectrometers like the VTPR were flown here, for example, as were many other experimental devices. Currently, the NIMBUS E Microwave Spectrometer, NEMS, is testing the potential of the microwave region for both temperature and humidity profiles of the atmosphere in the presence of clouds, which affects readings of the VTPR.

115

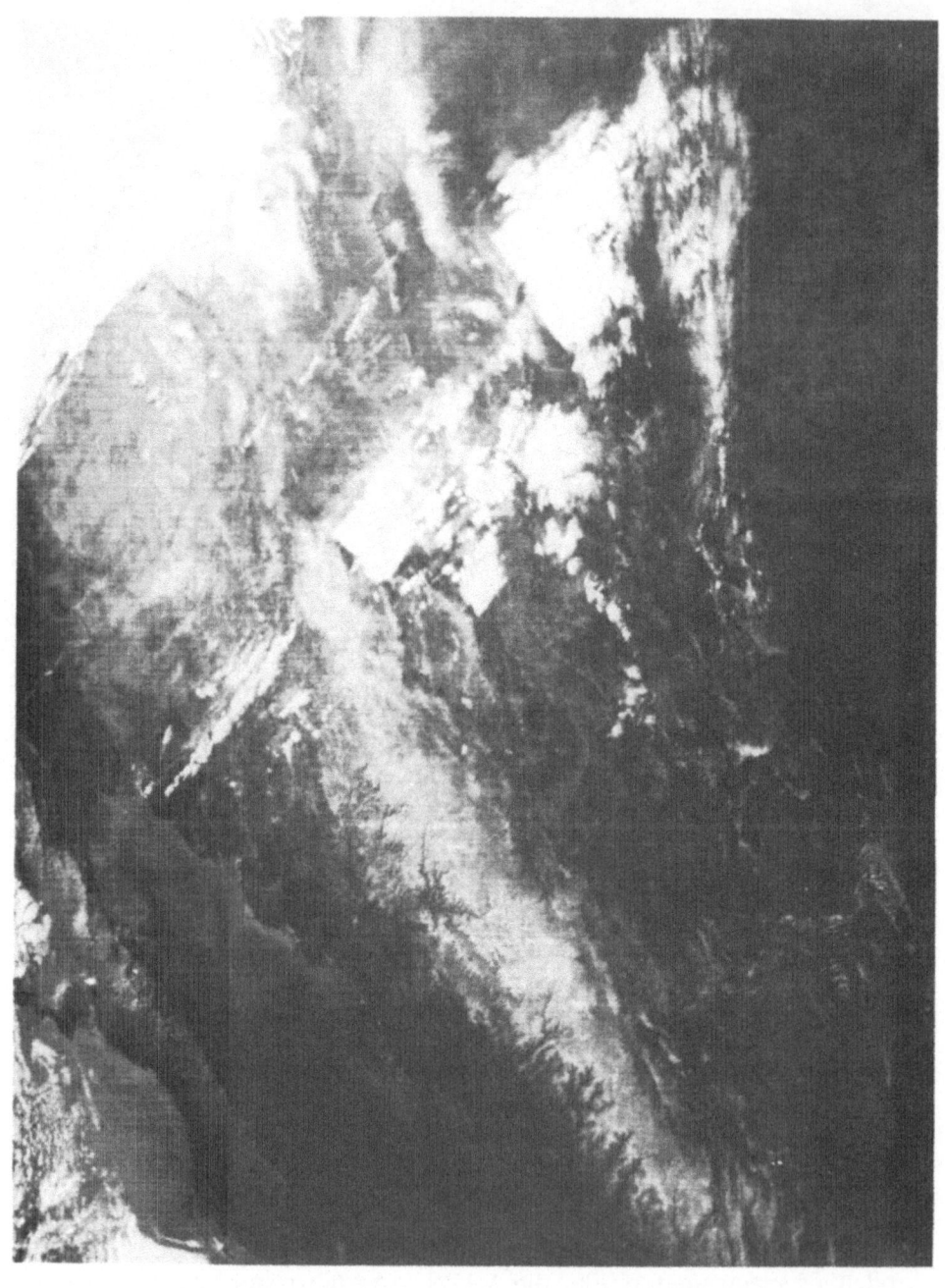

Plate 9-11. Thermal infrared image from the very high resolution radiometer of the NOAA-4 satellite.

116

Details of the NIMBUS sensors are given in a series of users' manuals, normally issued at about the time of launch of the satellite (39). Presently, Nimbus 6 is functioning nicely.

Access to NIMBUS data is by correspondence with the NIMBUS office, Goddard Space Flight Center, Greenbelt, Maryland 20771, U.S.A.

As a general rule, limited amounts of information are furnished to scientists on request, provided only that they have the means of analyzing the data that is requested. In addition, certain data are catalogued and printed; for example, a complete collection of THIR images has been printed and published in a series of manuals (33).

10. Geosynchronous Meteorological Satellites

The Geostationary Orbiting Earth Satellites, GOES (also called the Synchronous Meteorological Satellites, SMS, by NASA), are located in geosynchronous orbits. That is to say, they are located over the equator and rotate around the earth at the same velocity as the earth turns on its axis. Therefore, they appear to remain motionless in the sky, hanging at a nearly fixed point over the equator.

The first SMS/GOES satellite was launched in May of 1974, and more are launched periodically. Two are maintained in constant operation, one over Ecuador or Brazil, and the other over the Pacific Ocean; others are maintained in stand-by orbits in case one of these fails. Their primary sensor, an imaging radiometer, is called the Visible Infrared Spin Scan Radiometer, VISSR. This device takes whole-hemisphere images at about half-hour intervals. Images are taken in both visible and thermal-infrared radiation, with 2 km resolution in the visible and about 8 in the infrared (25). Photographic images, such as Plate 9-12, and images on computer tapes, for any hour of any day, are available for purchase from the Environmental Data Service, National Oceanic and Atmospheric Administration, Washington, D.C. 20233, U.S.A.

The two SMS/GOES satellites cover only the western hemisphere and parts of the Pacific Ocean. Three other satellites, roughly similar in capabilities, have been launched for other parts of the world: Meteosat, over Africa, by the European Space Agency; GMS in sight of eastern Asia by the Japanese government, and another SMS/GOES satellite over India, by the U.S., for central Asia. The satellites were first used together in FGGE, the First Global GARP Experiment (Global Atmospheric Research Program) in 1978. Together they furnish complete global coverage. Data are also said to be available for unrestricted use, but details are not available at this writing.

11. Automatic Picture Transmission (APT)

One undisputed benefit of the space age, especially useful for remote areas, are APT transmissions. Whereas most data from satellites require large,

Plate 9-12. Whole-hemisphere image from SMS/GOES-1.

expensive stations, APT signals can be received on small stations which are
inexpensive enough for use of individuals or small organizations. Signals can
be received anywhere in the world, without permission of any one, for any
purpose whatsoever. (22, 32).

NOAA, the National Oceanic and Atmospheric Administration, has an
office which responds to the needs of all APT users, APT Coordinator,
NESS-NOAA, FOB-4 Suitland, Maryland 20233, U.S.A.

This office receives correspondence from all parts of the world and
responds to any modern language. However, its replies, as well as the
information notes it publishes, are all written in English. The information
presented here came from that office.

Construction of the simplest APT receiving station is said to cost around
$3000 using used parts, including a facsimile image output. Components for
a new minimal station with facsimile output probably cost more nearly
$10,000 or 15,000. Addition of an oscilloscope readout with polaroid

118

camera recording of the image will greatly increase the quality of the recorded image but will add about $2000 to the cost of the station. At the time of writing, however, prices are in flux.

More details are available from the APT office. The Coordinator is especially anxious to help amateurs or small groups of students establish stations from available parts. However, his office cannot furnish either equipment or funds.

The NOAA/ITOS/TIROS satellites are the only ones scheduled to transmit real-time images to APT stations in the near future, but other images and information are relayed by the SMS/GOES satellites. The operator of a simple APT receiver can receive three real-time images of his area from the NOAA satellite in current use, two by day (visible and infrared) and one by night (infrared). He can also receive a variety of other images which have been processed on land and relayed to him by use of the SMS/GOES satellites.

Table 9-2 presents a representative schedule of WEFAX (weather facsimile) transmission from GOES-EAST (the SMS/GOES satellite over South America). Most entries in the table concern specific images, 'Mosaics' being composite images from the NOAA satellite and 'Quads' being quadrants of the SMS/GOES image of the hemisphere. Most of these images contain political and rough political boundaries, electronically added to the satellite signal during processing on the ground before it is relayed to the operators of APT stations.

12. Obtaining Imagery from the EROS Data Center.

Probably images of all highlands of the world have been acquired by NASA at one time or another. All of these images are in the public domain, available to any person or organization at a reasonable price. They constitute an invaluable source of information on the highlands.

The central organization which handles all NASA imagery is the EROS Data Center, U.S. Geological Survey, Sioux Falls, South Dakota 57198, U.S.A. This one organization maintains copies of all NASA imagery of the earth, and has the responsibility for distributing them to the public. It serves the international public as well as citizens of the United States, and it recognizes no restrictions or secrecy in this task.

Recently, stations outside the United States have begun accumulating archives of Landsat data. In particular, the following organization maintains images collected from central South America: Instituto de Pesquisas Espaciais, Calxa Postal 01, S. Paulo, Brazil. In general, archives of this and other new stations are not duplicated in the EROS Data Center.

At the time of writing EROS may have one or more of the following products for any given highlands area:

Gemini, Apollo and Apollo–Soyuz photographs, mostly from hand-held Hasselblad cameras with 70 mm film.

Table 9-2. Representative WEFAX Transmissions: GOES-EAST WEFAX Schedule (as of August 1978).

Id	Time sent	Data description
GE01	0020 GMT	1800 GMT DMSP S.H. Polar Mosaic 80W-10E-110E IR
GE02	0050 GMT	0001 GMT GOES-E NE/SE Quads IR
GE03	0120 GMT	0001 GMT GOES-E NW/SW Quads IR
GE04	0150 GMT	0001 GMT GOES-E Tropical W/E IR
GE05	0250 GMT	1800 GMT DMSP S.H. Polar Mosiac 110E-170W-80W IR
GE06	0350 GMT	0300 GMT GOES-E NE/SE Quads IR
GE07	0420 GMT	0300 GMT GOES-E NW/SW Quads IR
GE08	0750 GMT	0300 GMT GOES-E Tropical W/E IR
GE09	0820 GMT	2115 GMT DMSP S.H. Polar Mosaic 80W-10E-110E IR
GE10	0850 GMT	2115 GMT DMSP S.H. Polar Mosaic 110E-170W-80W IR
GE11	0950 GMT	0900 GMT GOES-E NE/SE Quads IR
GE12	1020 GMT	0900 GMT GOES-E NW/SW Quads IR
GE13	1050 GMT	0900 GMT GOES-E Tropical W/E IR
GE14	1120 GMT	0945 GMT GOES-W NW/SW Quads IR
GE15	1150 GMT	TBUS and Operational Messages
GE16	1250 GMT	1200 GMT GOES-E NE/SE Quads IR
GE17	1320 GMT	1200 GMT GOES-E NW/SW Quads IR
GE18	1350 GMT	1200 GMT GOES-E NE/SE Quads VIS
GE19	1420 GMT	1200 GMT GOES-E Tropical W/E IR
GE20	1450 GMT	1245 GMT GOES-W NW/SW Quads IR
GE21	1550 GMT	1500 GMT GOES-E NE/SE Quads IR
GE22	1620 GMT	1500 GMT GOES-E NW/SW Quads IR
GE23	1650 GMT	1500 GMT GOES-E NE/SE Quads VIS
GE24	1720 GMT	1500 GMT GOES-E Tropical W/E VIS
GE25	1750 GMT	1500 GMT GOES-E NW/SW Quads VIS
GE26	1820 GMT	1745 GMT GOES-W NW/SW Quads VIS
GE27	1950 GMT	1800 GMT GOES-E NE/SE Quads IR
GE28	1920 GMT	1800 GMT GOES-E NW/SW Quads IR
GE29	1950 GMT	1800 GMT GOES-E NW/SW Quads VIS
GE30	2020 GMT	1800 GMT GOES-E NE/SE Quads VIS
GE31	2120 GMT	1530 GMT DMSP N.H. Polar Mosaic 80W-70E-110E VIS
GE32	2150 GMT	2100 GMT GOES-E NE/SE Quads IR
GE33	2220 GMT	2100 GMT GOES-E NW/SW Quads IR
GE34	2250 GMT	2030 GMT DMSP NH Polar Mosaic 110E-170W-80W VIS
GE35	2350 GMT	2100 GMT GOES-E NE/SE Quads VIS

Skylab photographs, on 35 mm or 70 mm film from hand-held Nikon or Hasselblad cameras. Vertical multispectral photographs from Skylab, taken with the S-190A system on 70 mm film in six spectral bands (that is six images of each scene). Vertical high resolution photographs from Skylab taken with the S190B Earth Terrain Camera on 5-inch film, mostly for areas in the United States. Multispectral scanner images from Skylab on computer tape (require computer facilities and special programs for use). Vertical images from the Landsat satellites, 100 miles square from the MSS, available as black and white photographically reproduced images or computer tapes. Images were taken in the following bands: 0.5 to 0.6 (green), 0.6 to

0.7 (red), 0.7 to 0.8 (near-infrared), and 0.8 to 1.1 μm (near-infrared) and some thermal-infrared images from Landsat-3.

Imagery from NASA aircraft overflights, normally only for areas within the United States, including some of the following: black and white photography, colour photography, colour infrared photography, multispectral photography, multispectral scanner data on computer tapes, side looking airborne radar data on computer tapes, and thermal infrared scanner data on tapes. Other aircraft imagery of the United States from the U.S. Geological Survey, Army Mapping Serice, Navy, Air Force, Forest Service, Bureau of Reclamation, Bureau of Land Management and others.

When possible, an inquiry to EROS by phone or letter should state the type of imagery that would be most useful. It should list the country and the area within the country, including some geographical description. Since most searches will be made by computer, and most data are coded by geographical coordinates, it would be preferable to list the longitude and latitude of the area of interest. It is often helpful to the EROS specialists if the request specifies the uses to be made of the images. In the case of tapes, it is vital to specify the formates acceptable to the computers that will be used to process the data. Finally, since many otherwise useful images are invalidated by clouds, percentage limitations on cloud cover should be specified.

13. Independent Ways of Identifying Gemini, Apollo, Apollo–Soyuz, Skylab and Landsat Imagery

In most cases, one should depend on the specialists at EROS for identifying images of given areas that might be of use. However, in many cases, separate catalogs of images have been published. For the person who prefers to search catalogs, this section presents data that may be of use.

References (9) and (10) list some magnificent publications of NASA concerning the Gemini flights. Both contain about 250 reproductions in colour of Gemini photographs of the earth, and both are very reasonable priced. They also list all the photographs from these flights in an appendix. Listings include details on cameras, films, and filters, but lack many indices of utility of data such as cloud cover.

Reference (14) is another useful publication, reproducing images from Gemini flights as well as many unmanned flights. It identifies images only by mission, and it does not present listings of photographs. However, since there are relatively few images from the Gemini flights, most specialists will be able to identify a photograph if given only it's mission and a description of the scene.

The Technology Applications Center of the University of New Mexico publishes catalogs of pictures which it offers for sale. To date it has published reference (15) on Gemini flights and (6) on Apollo flights. In practice, the latter has been the only easily accessible index to the Apollo photographs. The center has plans to publish a catalog of Skylab photographs also.

The Lockheed Electronics Company, through its Mapping Sciences Department, was asked to prepare a computer retrieval system to allow automatic identification of Apollo and Gemini imagery. The retrieval system is no longer used extensively, but a printout of imagery for all regions of the world was published as an internal document for NASA (23). Since its aim was to describe the data base (that is, the contents of the memory queried by the computer program) it is not especially easy to use. However, it is the only comprehensive listing of Apollo and Gemini photographs which includes sufficient useful information on each photograph. Typical sections include, for example, all photographs of 'COLO, E. UTAH–WYOMING'; 'AFGANISTAN-W. PAKISTAN'; and 'PERU-ECUADOR'.

Monthly catalogs are published on paper, microfilm or microfiche, listing the Landsat images added to NASA's collection. These catalogs are at times available through the EROS Data Center. Periodically, cumulative catalogs are issued. Images are cataloged serially and by geographical coordinates. Some illustrations are presented to facilitate scanning of the data. There are two series, one for the United States and the other for the rest of the world, but foreign areas near United States boundaries are included in the United States listing.

Several other references may also be found to be useful. 'México desde el Espacio', a monograph to be published by the Comisión Ejecutivo del Territorio Nacional, contains a comprehensive list of all space imagery of Mexico (19). A similar publication, 'Bolivia from Space', has also been published for that country (16). Many research articles, such as ones concerning salt flats (26) or tectonics (8) list images that have proved useful; these are often valuable sources of identifications of useful images. Finally, the general references on space photography may be quite useful, furnishing representative photographs (9, 10), index maps, and other useful references. In addition, there are map indices available for Gemini (18) and Apollo (5) photographs, and a map index for some Skylab photographs is appended to a description of Skylab earth resources experiments (40). Master lists of Gemini (17) and Apollo (20) photographs have also been published.

14. Some Comments

Nations rich enough to develop space technology have little pressing need for it. Nations which truly need it are mostly too poor to develop it, and they are often nearly completely ignorant of its potential. This supreme irony of the space age is especially apparent in the highlands which are mostly in underdeveloped countries.

The United States and Russia, the two most advanced nations in space technology, have both expressed their desire to exploit peaceful applications of space technology. Both have offered to share their technology with the poorer nations. Nevertheless, the availability of LANDSAT images and the

APT system are perhaps the most widespread benefits yet to come to the underdeveloped world from the space technology of these two countries.

15. References

Many references on space technology are not available in the traditional way through journal articles or books. However, virtually all are available for the cost of copying through the National Technical Information Service, NTIS, U.S. Department of Commerce, Box 1551, Springfield, Virginia 22151, U.S.A.

An order for such materials should include, if possible, the Scientific and Technical Reports (STAR) number, which is always of the form N71-61313, or the International Aerospace Abstracts (IAA) number, which is always of the similar form A72-46212.

NASA has established an organization to disseminate technical information on space enterprises to the scientific community in the United States: the National Space Science Data Center, Goddard Space Flight Center, 601.4 Greenbelt, Maryland 20771, U.S.A. For scientists outside the United States the following organization serves the same purpose and is physically located in the same area: World Data Center A for Rockets and Satellites, Goddard Space Flight Center, 601, Greenbelt, Maryland 20771, U.S.A. These organizations supply single copies of references on aerospace technology, with special emphasis on rocket and satellite experiments. Especially valuable are their 'Reports on active and Planned Spacecraft and Experiments', issued annually with quarterly supplements.

References

1. Adams, J. S. *et al.* 1971. Development of Remote Methods for Obtaining Soil Information (Gamma Ray Detectors and Methods for Remote Sensing of Rocks and Soils), Report, Rice University, Houston 72N24454
2. Allied Research Associates, Inc., Meteorological Data Catalog for the Applications Technology Satellite, ATS-1 User's Guide, Volume 1, NSA-GSFC, Greenbelt, Maryland (1967) N71-11601 and N68-37369; ATS III User's Guide and Data Catalog, including ATS-I Data Catalog (second part) and ATS-II Summary (1969) N71-11602; see also N71-11603, N71-11604, N71-26621, and N72-33849.
3. Allison, L. J., E. B. Rodgers, T. T. Wilheit, & R. Wexler. 1975. A Multisensor Analysis of Nimbus 5 Data Recorded on January 22, 1973, NASA Technical Note D-7911, Goddard Space Flight Center, National Aeronautics and Space Administration, Greenbelt 74N22115.
4. Amaral, G. 1974. Remote Sensing Applications for Geology and Mineral Resources in the Brazilian Amazon Region, in Seminar on Space Applications of Direct Interest to Developing Countries, Sao Jose dos Campos, Brasil 75A22541.
5. Apollo Earth Photographs Index Maps, 1970. Apollo Missions 6, 7 and 9, NASA REDAF-51-00301, Houston.
6. Apollo Synoptic Photography Catalog, Technology Application Center, University of New Mexico, Albuquerque 87106.
7. Arnold, H. J. P. 1974. The Camera's Role in Remote Sensing from Space. Fundamentals of Remote Sensing. Proceedings of the First Technical Session, London 75A30836.
8. Carter, W. D. 1974. Tectolinear Interpretation of an ERTS-1 Mosaic, La Paz Area, Southwest Bolivia, Southeast Peru, and Northern Chile, presented at the Committee on Space Research, Seventeenth Plenary Meeting, San Jose dos Campos, Brazil.
9. 'Earth Photographs from Gemini III, IV and V,' NASA SP-129, U.S. Government Printing Office, Washington (1967).
10. 'Earth Photographs from Gemini VI through XIII', NASA SP-171, U.S. Government Printing Office, Washington (1968) 69N28909

11. Entres, S. L. 1974. Fundamentals of Remote Sensing of the Earth. Fundamentals of Remote Sensing, Proceedings of the First Technical Session, London A7530831.
12. Environmental Satellite Imagery (key to Meteorological Records Documentation No. 5. 4), National Oceanic and Atmospheric Administration, U.S. Department of Commerce (monthly Publication), Washington.
13. Estes, John E. & Leslie W. Senger. 1974. Remote Sensing, Techniques for Environmental Analysis', Hamilton, Santa Barbara 74A25105
14. "Exploring Space with a Camera', NASA SP-168, U.S. Government Printing Office, Washington (1968) 68N34870.
15. Gemini Synoptic Photography Catalog, Technology Application Center, University of New Mexico, Albuquerque 87106.
16. Giddings, L. E. 1977. "Bolivia Desde el Espacio', National Technical Information Service, PB262889.
17. Giddings, L. E. 1977. 'Gemini Photographs of the World', Complete Index. NASA-JSC-12875. N77-27472.
18. Giddings, L. E. 1975. Index Maps for Gemini Earth Photography, NASA-JSC-09581 N75-28502.
19. Giddings, L. E. 1978. Mexico Desde el Espacio, Comision Ejecutivo del Territorio Nacional, Mexico City.
20. Giddings, L. E. 1977. Near Earth Photographs from the Apollo Missions and the Apollo-Soyuz Test Project NASA-12947. In four parts N78-17436 through 17439.
21. Grant, K. A. 1974. Side looking Radar Systems and their Potential Application to Earth Resources Surveys; Basic Physics and Technology, Revue Scientifique et Technique CECLE/CERS, vol. 6, pp. 117–136 74A42863
22. Ground Systems for Receiving, Analyzing, and Disseminating Earth Resources Satellite data, Working Group One, Committee Application Satellites of the International Astronautical Federation, Paris. (1974).
23. Hixon, S. B. 1970. 'Technical Working Paper: Photographic Data Computer Program: GEMSORT', LEC/HASD TSP-70-12, rev. A, prepared for Mapping Sciences Laboratory, NASA, MSC, Houston 74N71532
24. Holz, Robert, 1972. 'The Surveillant Science: Remote Sensing of the environment', Houghton Mifflin, Baston.
25. Hussey, W. John. 1974. The Geostationary Environmental Satellite System, 490–497, Eascon.
26. Johnson, William R. & Dean R. Norris. 1977. A multispectral Analysis of the Interface between the Brazil and Falkland Currents from Skylab, Remote Sensing of Environment 6, 271.
27. Klass, Phillip J. 1971. 'Secret Sentries in Space', Random House, N.Y.
28. Kodak Data for Aerial Photography, Kodak Booklet M29 (general reference) Kodak Filters for Scientific and Technical Uses, Booklet B-3; Applied Infrared Photography, Booklet M28; The Seventh Here's How, p. 1–8 of Kodak Photo Information Book AE90. All are available from Department 454, Eastman Kodak company, Rochester, N.Y. 14650.
29. Lindenbaum, J. 1972. 'Remote Sensing Analysis: A Basic Preparation', The Laboratory for Applications of Remote Sensing, LARS, Information Note 110474, 2nd ed. This is a fine introduction to numerical remote sensing. No STAR Number.
30. McMillan, L. M., et al. 1973. Satellite Infrared Soundings from NOAA Spacecraft, NOAA NESS 65, National Oceanic and Atmospheric Administration, U.S. Department of Commerce, Washington N73-32291; see also Hayden, C. M., The Use of the Radiosonde in Deriving Temperature Soundings from NIMBUS AND NOAA Satellite Data NOAA TM NESS 76, N77-13622.
31. Morris, D. B. 1974. Non-Imaging Remote Sensing Systems, in Fundamentals of Remote Sensing, Proceedings of the First Technical Session, p. 107–133, London 75A30835.
32. NASA, 'Construction of an APT Station for TIROS N' (exact title unknown), NASA (1978); see also 'Real Time (APT/HRPT/BEACON) Data Systems for the TIROS-N

124

Spacecraft Series (1976)', W. J. Hussey, The TIRON-N Polar Orbiting Environmental Satellite System, and others from the ATP Coordinator.

33. Nimbus 5 Data Catalog, vol. 11, containing data from 1 August through 30 September, 1974, LANDSAT/Nimbus Project, Goddard Space Flight Center, Greenbelt, Maryland (1975).

34. 'Proceedings of the Symposium on Potential Application of Remote Sensing to Economic Development in Developing Countries', Washington (1970) 72N71227. See also 'Proceedings of the Seminar on Space Applications of Direct Interest to Developing Countries', San Jose dos Campos, Brazil (1974) 75A22526 *et seq.*

35. Rudd, Robert, 1974. 'Remote Sensing, a Better View', Duxbury Press.

36. Scherz, J. P., A. R. Stevens, & C. R. Belak 1970. 'An Introduction to Remote Sensing for Environmental Monitoring', University of Wisconsin 72N24401.

37. Schmugge, T., P. Gloersen, T. Wilheit, & G. Geiger. 1974. Remote Sensing of Soil Moisture with Microwave Radiometers, *J. geophys. Res.*, 79: 317–323.

38. Schwalb, A. 1972. Modified Version of the Improved TIROS Operational Satellite (ITOS D-G), NESS 35, National Oceanic and Atmospheric Administration, U.S. Department of Commerce, Washington.

39. Sissala, John E. 1975. The Nimbus 6 User's Guide, The LANDSAT/Nimbus Project, Goddard Space Flight Center, National Aeronautics and Space Administration, Washington.

40. Skylab Earth Resources Data Handbook, JSC-09016, NASA-JSC, Houston 1976.

41. 'The University of Michigan Notes for a Program of Study in Remote sensing of Earth Resources', Michigan Univ., Ann Arbor (1968) 72N70345.

42. Willekens, A. J. & J. H. Breeman. 1975. Remote Sensing from Aircraft. In Proceedings, Eighth International Aerospace Instrumentation Symposium, London 75A28776.

125

The vegetation of highlands

M. S. Mani

1. Introduction

The highland vegetation is remarkable for the absence of trees, large tree-like shrubs, lianas, climbers, etc. Shrubs are generally confined to elevations immediately above the timberline or grow only in certain specially favourable localities at higher elevations where also they tend to become greatly dwarfed. While true trees are absent, some interesting plants of certain tropical and equatorial mountains are arborescent. The giant and the arborescent *Senecio johnstoni* and *S. kilimanjara* are, for example, characteristic of the vegetation between 3000 and 4800 m on Mt. Kilimanjaro. The giant *Lobelia* grows at an elevation of 3600 m on Mt. Kenya. The paramoflora of the Andes has the giant Compositae *Espeletia*, which grows to 9 m height and bears at the tip a tuft of hairy leaves. The Puna plant of Andes *Puva raimondii* in Peru and Bolivia grows to 3–5 m height. Such giant and arborescent species may not be confused with trees. The highlands are basically treeless country.

All true highland plants grow exclusively at high elevations and are integral parts of the high altitude ecosystem (26). The differentiation and evolution of these plants are events contemporaneous to the uplift of the mountains and they are thus autochthonous to high altitudes. They do not represent accidental colonizations by plants of the lowland biota. The high altitude plants may be recognized by certain peculiarities, which can be described as part of their specializations for the high altitude environment These peculiarities distinguish them from other related types in the lowlands. Such peculiarities are superimposed on taxonomic and other characters, so that the general physiognomy of the plants of the ecosystem is quite characteristic. Some of the peculiarities of high altitude plants in general have been recently described (25, 26).

Broadly speaking, the high altitude plants are essentially cryophytes and partly mesophytes or xerophytes. Plants grow at extreme altitudes, where vertebrates and most other animals, except perhaps insects, cannot normally survive because of the inadequacy of oxygen. Plants occur, for example, at much higher altitudes than animals on the Himalaya (26). The highest altitude at which an Angiosperm, *Christolea himalayensis*, has been found in the world is 6300 m on Mt. Kamet (34, 35). Similarly cushions of flowering *Stellaria decumbens* were observed by Swan (37) at an elevation of 6140 m on Mt. Makalu. Other Angiosperms, flourish at elevations above 5500 m and almost up to 6000 m in the Himalaya (26). The highest elevation at which a Pteridophyte, *Cystopteris fragilis*, is known at present on the Himalaya is 5500 m (26). Gymnosperms like *Ephedra* grow at elevations of 5200 m in the Himalaya. Moss and lichens flourish in great abundance, however, at much higher elevations than 6300 m on the Himalaya. A number of interesting lichens, like the yellow coloured incrustations of *Usnea*, Parmeriae and Gyrophorae, have been photographed by me on barren rock surface at elevations of 6300–6600 m on the Himalaya, where the intense summer insolation on the glacier ice produces a thin film of oozing moisture for brief periods. Reports of fungi and yeast in soil at elevations of 7600–8400 m on Mt. Everest may also be found in literature (37).

The abrupt fall in abundance at the timberline and the progressive thinning out with increase in altitude, observed in other organisms, do not hold good for the high altitude plants (25). The fall in abundance with altitude becomes conspicuous only at elevations close to the permanent snowline. Almost up to elevations of 4500–5000 m there is scarcely any loss either in the diversity or in the abundance of individuals. Many genera and species characteristic of the upper reaches of the forest naturally tend to disappear above the timberline, but there is a very pronounced replacement by perhaps even larger numbers of genera and species that are increasingly characteristic of the higher elevations. Further, many of the subalpine-zone plants tend to be secondarily vegetatively modified for the prevailing conditions, which do not seem to exclude them altogether from colonizing the higher elevations. These include *Allardia glabra*, *Parrya lanuginosa*, *Pegophyton scapiflorum*, *Gentiana urnula*, *Selinum corticoides*, *Delphinium brunonianum*, etc. The reader will find interesting discussion of the altitudinal limits of plants in Mani (26) Webster (47) and Zimmermann (48).

Considerable confusion seems to prevail at present regarding the highland vegetation. The high altitude and high boreal vegetations are widely assumed to be similar. The apparent similarity lies merely on the general cryophyte character, developed in correlation with the atmospheric cold, but considered in total isolation from every other environmental factor that operates only at high altitudes and which are wholly absent in the boreal tundra and arctic regions. It must not, however, be overlooked that the boreal vegetation is composed basically of only typical lowland plants, which have secondarily become specialized to the prevailing atmospheric cold. None of them is exposed to the chain effects of the semi-attenuation of atmosphere as in the high altitude environment. The high altitude areas on continental mountains are also often classed, together with boreal latitudes, as cold desert, primarily because of dominance of xerophyte characters. It must be emphasized here that the origin and development of xerophytic characters of plants differ fundamentally in the hot desert, at high altitudes on mountains and in the high boreal latitudes of the tundra and arctic regions (26).

The high altitude vegetation develops xerophytic characters primarily as a reaction to the atmospheric aridity, that of the tundra and arctic areas as a reaction to inadequate soil moisture and the vegetation of the hot desert as reaction to all-round lack of water, both in the soil and in the air (25, 26). The high altitude xerophytes have many distinctive characters that separate them from the lowland boreal and desert xerophytes. They differ from the

128

lowland xerophytes not only in their genesis, but also in their general habits and appearance. The physiognomy of mountain xerophytes is distinctive. Korovin (22) and Kashkarov *et al.* (20) have discussed in some detail the salient peculiarities of high altitude xerophytes.

From the general dominance of semi-xerophytes and xerophytes, it must not, however, be concluded that all highland plants are xerophytes. Numerous species are extremely delicate herbaceous plants, with unbelievably tender leaves and flowers that wilt rapidly within a few minutes of being uprooted. There is in reality nearly every gradation from the strictly hygrophile to arid steppes type of species: (1) delicate hygrobiont annual herbs, (2) delicate hygrophile herbs which have the capacity of becoming dried up and of reviving when remoistened by melt-water, (3) semi-hygrophiles (mesophytes) of grassy meadows, (4) xerophiles and (5) xerobionts (euxerophytes).

2. Characters of Highland Plants

It is perhaps not possible to single out any character as exclusively high altitude feature. A combination of, however, structural and physiological peculiarities serves to distinguish them. These peculiarities in their combination may be considered as the high altitude specialization and correlated with the combination of ecological factors specifically operating in the highland ecosystem. Some of these factors are evidently low temperature and aridity of atmosphere, intensity and duration of sunlight, ultraviolet intensity, etc. The action of these factors and the reactions of plants to them are only very incompletely undersood at present. The relation between photosynthesis and daylength in highlands was studied by Bemberg (1), light intensity, temperature and photosynthesis by Moser (28), Pisek *et al.* (32) Tranquillini (38, 39, 40) and others, wind and photosynthesis Caldwell (8, 9). The effects of ultraviolet on growth of highland plants were studied by Brodführer (3) and Caldwell (7). Isolated studies on certain factors like photoperiodism, ultraviolet radiation, etc. on growth, photosynthesis and other plant reactions, published so far, would however seem to indicate the existence of a most complex relationship between every aspect of plant structure, habit and physiology and the conditions prevailing in the highland ecosystem. A detailed discussion of these problems lies outside our scope.

Many high altitude plants are perennials that require several years to attain maturity. Although the plants continue to grow for several years, they remain characteristically low and spread close to the ground. Inspite of their small size, most of the high altitude plants are of great age. In the lowlands there is a very marked dominance of aerial perennial parts of the plants, but in the high altitude ecosystem the aerial parts of plants are not generally perennial. The dominant perennials of the highland ecosystem have annual aerial parts that wither away at the end of each growing season. They have on the other hand perennial underground stems, rhizomes, stolons, rootstocks, creeping and often also woody, long rooting runners, generally covered by thick layers of old leaf sheaths, expanded bases of petioles, etc., and sending out fibrous roots. These sprout forth annually, during the growing seasons, into prostrate or decumbent, or erect and dwarf or tufted aerial shoots that wither away after having set seeds.

There are a number of species in which the aerial parts do not wither away at the end of the growing season but persist as hard, much branched cushions and polsters, though lacking underground persistent stems, root-stock, etc. They remain buried under snow during winter. When the snow melts in summer, their buds reawaken and unfold new foliage and the plant resumes its growth, producing more and more branches, year after year, to give rise to hemispherical or dome-shaped growths. The perennials consti-tute the bulk of the cushion, spreading mats and polster plants of densely tufted, semi-dome or dome-shaped dwarf growths, hardly 30–40 cm tall, but usually only 15 cm high above the ground. Whether perennials or annuals, highland plants are predominantly dwarfs. The remarkable hemispherical matted tufts of *Arenaria* and dome-shaped cushion of *Thylacospermum* from the Himalaya and Pamirs are recognized even from a distance.

In a number of plants the internodes are greatly shortened so that the leaves are largely or wholly all radical. The cauline leaves, if any, are generally very small and few. In such extreme cases the leaves are greatly crowded together as, for example, in *Acantholimon* or whorled as in *Pedicularis* or imbricate and even rosulate as in *Alchemilla, Sempervivum, Sedum, Androsace, Thylacospermum, Cassiope, Morina* and *Diapensia*. Flat-tened rosettes of *Werneria* and *Nototriche*, with stout tap-roots, grow close to the ground on the Andes. In the Puna Andes, at elevations of 4000 m and above in Peru and Bolivia, the remarkable rosette plant *Puva raimondii*, bears inflorescence axis 3–5 m high. Ground rosettes of *Plantago, Viola* and *Calycera* are described from the Andes (46).

The shortening and suppression of internodes and the formation of rosettes are common characters of the alpine-zone plants on nearly all high mountains, as well as in the subarctic, arctic and sub-antarctic areas. The underlying factors of the rosette formation in these different ecosystems are complex and different. Rosettes may be explained by very low growth rate, resulting in shortening of the internodes. Intensity of sunlight perhaps inhibits the elongation of stem at high altitude – the long summer daytime of the high altitude and the 'midnight' sun of the high latitudes may be perhaps important factors in suppressing internodes. The temperature regime and retardation of the flow of sap over long distances of internodes, because of excessive atmospheric cold and snow, must not also be overlooked. Rosette formation, though common to the high altitude and high latitudes, may actually be very different responses. All rosettes at high altitude close tightly into compact buds, resembling a miniature cabbage, at night and reopen only after the sunrise. The central growing point, within the rosette, remains always closely and tightly ensheathed by the young leaves, which also continue to remain folded until the scape pushes upwards as it grows. The closing of rosettes at night is apparently related to fall in air temperature.

Many species are devoid of aerial stem at extreme high elevations. The absence of an aerial stem and the fact that all leaves are radical are conditions correlated with the blade being narrowed at base, broader apically and showing a strong tendency to be spatulate, obovate or oblanceolate and pinnatifid or pinnatisect as in Cruciferae, many Compositae and others. The generally great physiological age of foliage formation probably also would explain, at least partly, the predominance of dentate, lobed and pinnatisect conditions of the leaves at high elevations.

The inflorescence of rosette plants is generally scapigerous. The scapes are nearly always leafless and perhaps represent the only erect part of an otherwise low-growing plant. The peduncles and pedicels may also be conspicuously shortened and many plants have their flowers in corymbs or in shortly peduncled heads as in *Oxytropis*, *Trifolium*, etc. The same tendency should explain the dominance of compound umbels of Umbelliferae, the compound heads of the Compositae being gathered together is semi-umbellate or corymbose clusters. Massed spikes, verticillate bunches, short and compound umbels, and corymbose heads are the more common types of inflorescence at higher elevations. Single flowers and loose panicles may only be found in sheltered localities, where the general conditions do not appear to be so extreme. The massing and clustering of flowers at the tip of a generally leafless scape are perhaps parts of the same specializations and must be closely correlated with the formation of foliage rosettes. The rapid development of the inflorescence may perhaps inhibit the normal elongation of internodes and in this way indirectly favour the rosette formation. At any rate, the leaf rosettes, the long leafless scapes and the massing of flowers seem to be all interlinked characters, apparently related to the conditions prevailing at ground level in the high altitude ecosystem.

The foliage of high altitude plants presents a number of interesting specializations. There is nearly every gradation from the most delicate to the rather heavily coriaceous foliage, depending mainly on the special habitat. The preponderance of small-sized and coriaceous leaves is widely believed to be the result of action of ultraviolet. The high altitude flora of the Andes is characterized by the presence of a number of highly coriaceous and scaly-leaved shrubs like *Baccharis*, *Lepidophyllum*, *Loricaria*, etc. Other shrubs like *Befaria*, *Gaultheria*, *Hypericum*, etc. have small hard leaves. *Loricaria* has rolled leaves. Dense woolliness is also known to be produced in response to high intensity of ultraviolet. Some of the plants with rather conspicuously thick leaves, particularly at high elevations, belong to *Sedum*, *Sempervivum* and certain species of *Androsace*. Nearly all parts but particularly the leaves, are hoary, densely pubescent, villous, woolly or even cottony. For example, *Cheiranthus* is typically hoary and *Draba* is often tomentose. Among the pronouncedly woolly or silky haired plants are *Cimifuga*, *Meconopsis*, *Alyssum*, *Sisymbrium*, *Potentilla*, *Aster*, *Erigeron*, *Leontopodium*, *Tanacetum*, *Eritrichium*, etc. Profusely cottony plants include *Anaphalis*, *Helichrysum* and *Saussurea* among others. On account of the dense cottony hairs, *Saussurea gossypiphora* is used in Lahaul-Spiti as tinder by nomadic shepherds. Woolly herbs like *Pleurophyllum*, *Culicitum*, woolly polsters of *Raoulia*, *Haastia*, *Phycnophyllum*, *Mulinum*, *Azorella*, woolly

131

leaved *Helichrysum*, etc. are characteristic of the high altitude vegetation of the Andes (46). Woolly plants are characteristic of sub-arctic and sub-antarctic ecosystems also, but the varying underlying causes of woolliness evoke responses and secondary effects. Woolliness is associated with therma-insulation and retardation of evaporation. The entrapped air in between the long hairs has great significance in both these effects. Effective protection is also provided by dense cover of dried leaf bases, expanded petioles, tufted branches, persistent scaly parts, etc., that cover buds and perennial parts.

We may remark here on the general appearance of highland vegetation. While the general colour of the vegetation of the tundra and arctic north and even also of the Eastern Himalaya and the Alps is known to tend towards blue-green rather than green, high altitude vegetation of the Pamirs and the Northwest Himalaya is predominantly yellowish-green or has even an excess of brown (25, 26). These differences are explained as part of the specialization for effective reflection of up to 55% of the infrared of the solar radiation in the case of blue-green vegetation. The yellowish-green or brown vegetation reflects much less than 16% of the infrared of the sunrays (33). These peculiarities become meaningful if we recollect that there is an excess of blue and relatively high ultraviolet in the high altitude environment. The recent studies of Popova (33) have shown that the pigments of the high altitude plants do not differ basically from those of the lowland plants, but the chlorophyll content is significantly low in the high altitude species. Bukharin (5) has made the interesting observation that carotene content is also generally low in the vegetation of the Pamirs.

The flowers of the greatest majority of the high altitude plants are predominantly yellow, orange, pink red and white and only exceptionally blue or violet. The predominant colours of the flowers seem to be in some way related to the excessive blue and ultraviolet glare of high altitude. These are the colours which powerfully attract the flower-visiting diurnal Diptera and Lepidoptera and thus serve to favour cross-pollination. The vast majority of the insect-pollinated flowers in the forest and other lowland vegetation are visited by bees, especially the honeybees, Diptera and Lepidoptera. The lowland vegetation is remarkable for the great abundance of sweet-scented and night-blooming flowers, which are pollinated by crepuscular and nocturnal moths. At high altitude practically no insect is on its wings at night and the few active fliers are on their wings only during the hours of bright sunshine. The high altitude flowers bloom, therefore, only during the daytime and lack strong scents and nectaries. The high altitude flowers offer mostly pollen to the insect. Insect pollinated flowers are not also as numerous as the wind pollinated ones at extreme altitudes above 4000 m.

3. Some Typical Highland Plants

Nearly every group of plants like lichen, moss, liverwort, fern, Gymnosperm, Dicotyedon and Monocotyledon has specialized representatives in the highland ecosystem.

The dominant Phanerogams of high altitude are Cruciferae, Gramineae, Ranunculaceae, Compositae and Caryophyllaceae, followed by Leguminosae, Geraniaceae, Boraginaceae, Labiatae, Polygonaceae, Fumariaceae, Primulaceae, Scrophulariaceae, Liliaceae and Cyperaceae. The other families include Rosaceae, Saxifragaceae, Crassulacae, Umbelliferae, Caprifoliaceae, Ericaceae, Gentianaceae, Salicaceae and Juncaceae. The following examples may be considered as typical high altitude plants, the majority of which grow at elevations between 3600 and 6000, mainly in the Alpine-Himalayan System.

i. *Dicotyledons*

The Ranunculaceae are represented by *Aconitum, Adonis, Anemone, Aquilegia, Callianthemum, Caltha, Delphinium, Oxygraphis, Ranunculus, Thalictrum, Trollius*, etc. at elevations of 2000–5500 m. Most species are dwarf or stemless, much branched and tufted perennial herbs.

The Cruciferae are dominant in semi-arid continental highlands and are characterized often by hairy, tufted, dwarfed and much branched perennial herbs like *Alyssum, Arabia, Barbarea, Braya, Capsella, Cardamine, Cheiranthus, Chorispora, Christolea, Draba, Erysimum, Parrya, Sisymbrium*, etc. common at elevations of 3600–5500 m and often growing even above 6000 m on the Himalaya. The Caryophyllaceae are cushion-plants or densely tufted and hairy species of *Arenaria, Cerastium, Gypsophila, Lychnis, Sagina, Silene, Stellaria*, etc. growing often at elevation of 6000–6300 m. The remarkable hemispherical cushions of greatly tufted dwarf herb *Thylacospermum* at elevations of 4500–5500 m on the Himalaya belong to this family.

On grassy meadows at elevations of 3600–5500 m are found hairy or hoary and much branched tufted herbs, *Biebersteinia* and *Geranium*, with perennial and often stout underground rootstock, belonging to the family Geraniaceae.

The Leguminosae are mostly characteristic of arid steppes in highlands, though sometimes found even in grassy meadows. *Astragalus* and *Caragana* are dwarf shrubs at 3300–5500 m on the arid interior of the Himalaya and the Pamirs. *Oxytropis* is often a stemless, densely silky haired perennial herb, not more than 10 cm tall, growing from woody rootstock up to 5500 m on the Himalaya. *Thermopsis* is a perennial herb, with much branched aerial shoot arising from woody rootstock upto an elevation of 5500 m. *Trifolium* are prostrate herbs, which sometimes occur at 6000 m.

The common Rosaceae of the highland ecosystem include *Alchemilla* and *Potentilla* which are typically much branched spreading or tufted plants, growing upto 5500 m on the Himalaya.

Saxifragaceae. A number of species of *Saxifraga*, which are generally short stemmed or densely tufted perennial herbs, are common. *Saxifraga engleriana* grows at 5700 m on the Himalaya.

Crassulaceae. *Sedum* is typically a succulent perennial with rootstock, ending in a tuft of scales and tufted branches and mostly rosulate leaves at elevations between 3600 and 5500 m on the Himalaya and is also reported from the Abyssinian Highlands and from the Peruvian Andes.

Umbelliferae are common enough on highland grassy meadows, at elevations of 3900–4500 m, but some remarkable giant forms are characteristic of semi-arid areas. The common species belong to *Bupleurum, Heracleum, Pimpinella, Pleurospermum* and the densely hairy, stout-stemmed dwarf *Trachydium* is found upto 5500 m on the Himalaya.

The Compositae are characteristic of arid-frigid steppes and are generally dominant members of the vegetation at elevations of 3900–5500 m. Some of the typical forms are species of *Allardia, Artemisia, Aster, Centaurea, Cnicus, Crepis, Echinops, Erigeron, Gerbera, Helichrysum, Hieracium, Inula, Lactuca, Leontopodium, Saussurea, Scorzonera, Senecio, Taraxacum,* etc. *Allardia* is densely woolly, tufted aromatic dwarf herb *Aster* is usually a much branched, often tomentose dwarf herb, *Centaurea* is a tomentose or hoary herb and *Echinops* is a densely tomentose plant. *Erigeron* tends to become a tomentose and woolly dwarf at high elevations. *Helichrysum* is a characteristically woolly herb. *Inula* is stemless or has numerous dwarfed stems clumped together on the woody rootstock.

Leontopodium is a tufted woolly perennial herb found upto 5500 m on the Himalaya and common in the Pamirs. *Saussurea* species are generally stemless but sometimes also densely tufted dwarfs, usually highly tomentose or also densely cottony as in *S. gossypiphora* at 4500–5100 m on the Himalaya. *Senecio* are dwarf shrubs or much branched or spreading herbs, but sometimes arborescent as in *s. johnstoni* from Mt. Kilimanjaro.

Ericaceae are particularly characteristic of tropical highland regions, especially East Africa and South America and include *Gaultheria* a much branched hirsute shrub, *Rhododendron* spp., etc.

The Primulaceae are generally characteristic of grassy meadows and include *Androsace, Lysimachia* and *Primula.* Some of them grow at the snowline at elevations of 5100 m on the Himalaya (Plate 10.1,2).

The Gentianaceae are represented by interesting species of *Exacum, Gentiana* and *Swertia*, which are generally much branched dwarfs, common between 3000 m and 5100 m on the Himalaya.

Subarid steppes species of the Boraginaceae, found on the Himalaya and Pamirs, include *Arnebia, Cynoglossum* and *Echinospermum* at 3000–3900 m, *Eritrichum* and *Lindelofia* up to almost 5000 m, *Macrotomia* at 3000–4200 m, *Myosotis* found up to 5100 m, etc.

Scrophulariaceae are also generally found in semi-arid and rocky areas and include *Lagotis, Pedicularis, Picrorhiza, Veronica,* etc., which are generally common at elevations of 3300–5500 m on the Himalaya.

Many Labiatae are typical of cold, wind-blown, semi-arid steppes, at elevations of 3000–5000 m and include aromatic-herbs like *Calamintha, Dracocephalum, Elsholtzia, Hyssopus, Nepeta, Perovskia, Phlomis, Thymus,* etc.

Some of the minor families of dicots may now be mentioned. Chenopodiaceae with *Chenolea* a silky-villous or tomentose and much branched dwarf shrub at 3300–4200 m, the woolly or tomentose and densely branched dwarf *Eurotia* found up to 4500 m; Polygonaceae with steppes species of *Polygonum* occurring up to 5500 m; the spinescent and curiously stunted and densely branched shrub *Hippophaë* of the Elaeagnaceae.

ii. *Monocotyledons*

Liliaceae include the high altitude wild onions, *Allium* spp. scapigerous herbs with tufted and coated bulbs, often growing up to 5000 m and *Fritillaria* found at 3000–4000 m. *Gagea, Lloydia, Trillium*, etc. *Juncus* species with tufted stems and stout rootstock grow up to 4200 m. *Cobresia* is a perennial with short and sometimes woody rhizomes at 4900 m. *Carex* with about half a dozen species has clustered short stems at 3300–5000 m. The Gramineae are perhaps among the commonest high altitude plants on the Himalaya, with over a dozen peculiar genera like *Stipa, Deyeuxia, Deschampsia, Danthonia, Catabrosa, Poa, Glyceria, Festuca, Bromus, Agropyrum* and *Elymus*. The majority of the species are densely tufted, often also prostrate, perennials, which occur as high as 5500 m.

4. African Mountains

The high altitude vegetation of the African mountains is fundamentally different from that of the Himalaya. At elevations comparable to the treeless zone of the Pamirs and the Himalaya, many plants have peculiar arborescent habits on Mt. Kilimanjaro and Mt. Kenya (25). *Helichrysum* and the arborescent *Senecio johnstoni* grow, for example, at elevations of 3000–5200 m on Mt. Kilimanjaro. Immediately above the upper limits of the closed forest on the Equatorial mountains lies the so called lower moorland zone, up to an elevation of 3500 m and is characterized by *Erica, Philippia, Protia, Adenocarpus, Hypericum, Artemisia*, giant *Senecio kilimanjara*, sedges and grasses. In the upper moorland zone, the dominant plants comprise *Philippia jaegeri, Euryops dacrydioides, Helichrysum*, giant *Senecio cottoni*, sedges and grasses. At elevations of 3970 m and 4300 m, the vegetation is open and higher above lies alpine-desert species of tufted grasses, cushion plants, etc. Though Angiosperms are relatively sparse here, lichens and mosses like *Grimmia ovalis, Hypnum cupressiforme, Torula cavallii* and *Webera afrocauda* are common. Because of the diurnal (and not seasonal) frost, there is here practically no specific growing season, distinct from the non-growing season.

Lobelia keniensis, Senecio, Hagenia and *Festuca* occur at elevations of 4500 m on Mt. Kenya. The common mosses include *Andreaea cucullata, Bartramia afroithophylla, Brentelia subgnathophalea, Erimmea ovata, Rhacocarpus humboldtii* and *Rhacomitrium durum*. In the high alpine zone occur species of *Cerastium, Oreophytum, Tachydium, Senecio, Agrostis, Anthoxanthum, Festuca, Sagina, Arabis, Alchemilla, Swertia, Myosotis, Calamintha* and *Valeriana*.

135

Plate 10-1. 1. The high altitude region of the Himalaya—a treeless world of snow, ice and rock. 2. *Primula denticulata* in bloom at the edge of snow at an elevation of 4500 m on the Himalaya.

5. The Andes

The high altitude plants of the Andes present a number of peculiarities (25), remarkable for the number of highly coriaceous and scaly-leaved shrubs like *Baccharis, Lepidophyllum, Loricaria,* etc., some of which are characteristically resinous and aromatic and grow in bunched clusters and are often also giants. There is also an abundance of shrubs with small hard leaves like *Befaria, Gaultheria, Hypericum,* scaly and roll-leaved *Loricaria,* woolly-leaved *Helichrysum,* polster plants like *Azorella, Distichia, Plantago, Acichne* and *Oreobolus* and flattened rosettes close to the ground and with stout tap roots as in *Werneria* and *Nototriche.* The equatorial section of the high Andes is remarkable for the Paramo, rising up to elevations of 4800 m. The paramo is typically a treeless region, with cold, heavy clouds, storms and essentially diurnal-thermic climate. Immediately above the forest the typical paramo vegetation comprises an abundance of loosely scattered and open shrub of Ericaceae, *Hypericum,* Compositae like *Diplostephium, Senecio,* etc. Higher above this come tufted grasses *Stipa, Calamagrostis* and somewhat higher still in moist localities dominant polster plants of Compositae, Valerianaceae, Plantaginaceae, *Azorella,* etc. There are also extensive *Distichia* moors. The most remarkable of the paramo flora is *Espeletia,* a genus of the Compositae, with a stem rising often 9 m high and bearing at the tip a tuft of hairy leaves. The puna vegetation occurs at elevations of 4000 m and above in Peru and Bolivia and is characterized by *Stipa, Calamagrostis, Festuca, Puva raimondii,* a rosette plant bearing inflorescence axis 3–5 m high. The occurrence of the tola shrub, highly resinuous Compositae *Lepidophyllum quadrangulare,* with small scaly leaves, is another striking feature of the puna. The Sub-Antarctic area of the Andes is characterized by the high altitude plants like *Azorella, Colohanthus, Lycopodium saururus, Oreobolus, Donatia, Astelia, Gaimardia, Abrotanella, Arctiastrum, Lomaria,* woolly herbs like *Pleurophyllum, Culicitum,* woolly polster *Raoulia, Haastia* and *Pycnophyllum* and tussock grass.

6. References

1. Bamberg, S., W. Schwartz & W. Tranquillini. 1967. Influence of day-length on the photosynthetic capacity of stone-pine (*Pinus cembra*). *Ecology,* 48: 264–269.
1a. Baruch, Z. 1979. Elevational differentiation in *Espeletia schultzii* (compositae), a giant rosette plant of the Venezualen Paramos. *Ecology,* 60(1): 85–98.
2. Billings, W. D. 1975. Vegetation: Arctic and alpine vegetation. Plant adaptations to cold summer climate. In: Arctic and Alpine Environments, Ed. J. D. Ives & R. G. Barry. London: Methuen & Co.
3. Brodführer, V. 1955. Der Einfluss einer abgetuften Dosierung von Ultravioletter sonnenstrahlung auf das Wachstum der Pflanzen. *Planta,* 45: 1–56.
4. Brozoska, W. 1973. Stoffproduktion und Energiehaushalt von Nivalpflanzen. Ökosystemforschung (Ed. H. Ellenberg, pp. 2225–233).
5. Bukharin, P. D. 1961. Carotene levels of introduced plants grown at various altitudes. *Dokl. Akad. Nauk* SSSR (Bot.) 136 (1–6): 28–30.
6. Caberra, A. L. 1958. La vegetacion de la Puna Argentina. *Rev. Invest. agric. B. Aires,* 11 (4): 317–412.

7. Caldwell, M. M. 1968. Solar ultraviolet radiation as an ecological factor for alpine plants. *Ecol. Monogr.*, 38 (3): 243–268.

8. Caldwell, M. M. 1970. The wind regime at the surface of the vegetation layer above the timberline in the Central Alps. *Centralbl. Ges. Forstw.*, 87: 65–74.

9. Caldwell, M. M. 1970. The effect of wind on the stomatal aperture, photosynthesis and transpiration of *Rhododendron ferrugineum* and *Pinus cembra*. *Centralbl. Ges. Forstw.*, 87: 193–201.

10. Cuatrecases, J. 1958. Aspectos de la vegetacion natural de Colombia. *Rev. Acad. Colon. Cienc. exact. fis. nat.*, 10: 10.

11. Diels, L. 1934. Die Paramos der äquatorialen Hochanden. *Sitzb. preuss. Akad. Wiss.*, 57–58.

12. Engler, A. 1914. Über Herkunft, Alter und Verbreitung extremer xerothermer Pflanzen. *Sitzb. preuss. Akad. Wiss.*, 20.

13. Fedschenko, B. A. 1902. Ocherk k rastitelnoti Pamira Shungnana e Alaiya. Trud. S. Petersburgskovo estestvoispitalei, 33: 1.

14. Fedschenko, B. A. 1903–1909. Flore du Pamir. Acta horti Petropolitana, 21 (1903); Suppl. 1. 24 (1904); 2: 24 (1905); 3: 28 (1907); 4: 28 (1909).

15. Fedschenko, B. A. 1947. Botanicogeographicheskoe rainoni rovanie Pamira. Tezisi Dokladu II. Sbezda vesoyuzhnovo Geographicheskovo Obschestovo.

16. Garnes, W. W. & H. A. Allard. 1920. Effect of the relative length of day and night and other factors of the environment on growth and reproduction in plants. *J. agric. Res. Washington DC*, 18 (11): 553–606.

17. Golovkova, A. G. 1957. Visokigornaya rastitelnost Centralnova Tian-Shanya. *Sbornik Kirghiskovo Nauchno Issledovatleskovo Inst. Pedogogiki*. Frunze.

18. Golovkova, A. G. 1959. Rastitelnost Centralnovo Tian-Shanya. Kirghiz-Gosudarst. Universtitet, Kaf. Bot. Frunze, pp. 1–456.

19. Hauman, L. 1918. La vegetattion des hautes Cordilléres. *Ann. Soc. Bien. Aug.*, 86: 121–128.

20. Kashkharov, D. N., A. I. Zhukov & K. V. Stanyukovich. 1937. Kholodnaya Pustinya Centralnovo Tian-Shanya.

21. Khalikov, M. K. 1963. Prodolzhitelnost zimnego pokoya nekotorykh rastenii Pamira. *Tr. Pamiriskoi Biol. Sta.*, 1: 243–248.

22. Korovin, E. P. 1934/1962. Rastitelnosti Srednii Asii yuzhnovo Kazakhstana. 1. Tashkent (1934); II. Acad. Nauk Uzbek SSR, Tashkent, pp. 1-540 (1962).

23. Mani, M. S. 1962. Introduction to High Altitude Entomology. London: Methuen & Co. pp. 1–304.

24. Mani, M. S. 1968. Ecology and Biogeography of High Altitude Insects. The Hague: Dr W. Junk BV Publishers, pp. 1-527.

25. Mani, M. S. 1974. Fundamentals of High Altitude biology. New Delhi: Oxford & IBH. pp. 196.

26. Mani, M. S. 1978. Ecology and phytogeography of High Altitude Plants of the Northwest Himalaya: Introduction to High Altitude Botany. New Delhi: Oxford & IBH. London: Chapman & Hall pp. xii+205, pl. xxiv, figs. 49.

27. Mooney, H. A. & W. D. Billings. 1965. Effect of altitude on carbohydrate content on mountain plants. *Ecology*, 46 (5): 750–751.

28. Moser, W. 1973. Licht, Temperatur und Photosynthesis an der Station 'Hoher Nebelkegel' (3184 m). Ökosystemforschung (Ed. H. Ellenberg) pp. 203–223.

29. Müller, H. 1881. Die Alpenblumen, ihre Befruchtung durch Insekten und ihre Anpassungen dieselbe. Lepizig.

30. Penland, W. T. 1941. The alpine vegetation of the southern Rockies and Ecuadorian Andes. *Colorado Coll. Pub.*, (Gen. Ser.) 32: 230.

31. Pisek A. & E. Winkler. 1958. Assimilationsvermögen und Respiration der Fichte (*Picea excelsa*) in verschiedener Höhelage und der Zirbe an der alpinen Waldgrenze (*Pinus cembra*). *Planta*, 51: 518–543.

32. Pisek, A. & E. Winkler. 1959. Licht- und Temperaturabhängigkeit der CO_2-Assimilation von Fichte (*Picea excelsa* Link.), Zirbe (*Pinus cembra* L.) und Sonnenblume (*Helianthus*

annus L.) Planta, 53: 532–550.

33. Popova, I. A. O. pigmentakh listev pamirskikh rastenii. *Bot. Zhur.*, 43 (11): 1550–1561.

34. Rau, M. A. 1963. Illustrations of West Himalayan Flowering Plants. Calcutta: Botanical Survey of India. Special Pub. pp. 1–234.

35. Rau, M. A. Flora and vegetation of the Himalaya. In: Mani, M. S. 1973. Ecology and Biogeography in India. The Hague: Dr W. Junk. Ser. biol. pp. 247–280.

36. Shelter, S. W. 1964. Plants in the Arctic alpine environment. Ann. Rep. Smithsonian Inst., 4530: 473–497.

37. Swan, L. W. 1967. Alpine and aeolian regions of the world. In: Wright, H. E. Jr & W. H. Osburn. Arctic and Alpine Environments. Indiana Univ. Press. 3: 29–54.

38. Tranquillini, W. 1955. Die Bedeutung des Lichtes und der Temperatur für die Kohlensäure-Assimilation von *Pinus sembra* L. Jungwuchs an einen highalpinen Standort. *Planta*, 46: 154–178.

39. Tranquillini, W. 1957. Standortsklima, Wasserbalanz, und CO_2-Gaswechsel junger Zirben (*Pinus cembra* L.) an der alpinen Waldgrenze. *Planta*, 49: 612–661.

40. Tranquillini, W. 1967. Über die physiologischen Ursachen der Wald- und Baumgrenze. *Mittl. Forstl. Bundesvers. Anst.* Mariabrunn, 75: 456–487.

41. Troll, C. 1931. Die Landschaftsgürtel der tropischen Andes. *Verh. wiss. Abh.* 24 *deutsch. geogr. Tag. Danzig.* pp. 264–270.

42. Troll, C. 1959. Die tropische Gebirge, ihre dreidimensionale klimatische und pflanzengeographische Zonierung. *Bonn. geogr. Abh.*, 25.

43. Turner, H. 1958. Maximaltemperaturen oberflächennacher Bodenschichten an der alpinen Waldgrenze. *Wetter Leb.*. 10: 1–12.

44. Turner, H. 1970. Grundzüge der Hochgebirgsklimatologie. Die Welt der Alpen. Pinguin Verlag. Innsbruck.

45. Weber, H. 1958. Die Paramos von Costa Rica und ihre Pflanzengeographische Verkettung mit Hochanden Südamerikas. *Abh. Akad. wiss. lit.* Mainz, (Math-naturw. Klasse) 3.

46. Weberbauer, A. 1911. Die Pflanzenwelt der peruvianischen Anden. In: Vegetation der Erde. 12.

47. Webseter, G. L. 1961. The altitudinal limits of vascular plants. *Ecology*, 42: 587–590.

48. Zimmermann, A. 1953. The highest plants in the world. Mountain World, New York: Harper. pp. 130–136.

The animal life of highlands

M. S. Mani

1. Introduction

The animal life of highlands includes the hypsobiont species, the hypsophile visitors from lowland ecosystems, recent colonies from lower elevations and species accidentally or passively transported to highlands. The hypsobiont animals are mountain authchthonous forms. As all animal life evolved primarily in an environment of dense atmosphere near sea level, the hypsobiont species represent specialization to the environmental conditions of thin atmosphere, developed on site in the course of the uplift of the ground to high elevation. The hypsobiont animals differ from the lowland forms in their ecological, physiological and structural peculiarities, in their origin, evolution and history. They must be distinguished from the numerous species, which wander entirely accidentally to highland areas and from the hypsophiles, which migrate regularly and periodically to elevated regions. Then there are also a number of interesting species, which are passively transported by winds; they are all not permanent residents that breed and develop exclusively at high altitudes.

2. Peculiarities of Hypsobiont Animals

Broadly speaking, the hypsobiont animals fall into two major ecological groups: (i) the temperature oriented and (ii) the pressure oriented (more correctly oriented to the oxygen-deficiency) animals.

i. *Temperature Oriented Animals*

This group of hypsobiont animals are characterized by their primary reaction to the lowering of atmospheric temperature rather than directly to the lower oxygen content at high altitude. The capacity of poikilotherms of adjusting their vital activities to the temperatures of the microclimatic niches involves an effective utilization of the low temperature. The atmospheric cold of high altitude brings about a general lowering of basal metabolism and thus automatically reduces the respiratory function and oxygen requirement. As a chain reaction, this readjusts the animal to nearly every other environmental factor, which influences it at high altitude. By depressing the metabolic rate, the oxygen requirement is reduced, thus enabling the animal to survive and function normally at elevations, at which the oxygen deficiency of the air excludes most if not all the homoiotherm animals. It is thus found that poikilotherms occur at much higher elevations than the homoitherms. In the poikilotherm the adjustment to the atmospheric cold of high altitude favours considerable anaerobic functioning and at the same time retards various aerobic reactions and thus alters the acid-base balance, as adjustment to the prevailing extreme conditions. These changes result automatically in a more or less pronounced resistance to hypoxia and eliminate nearly all pathological changes induced by exposure to altitude. The reduced metabolism, the change in the acid-base balance and other related conditions lead to changes of neurohormonal and enzyme actions, thus indirectly influencing dormancy and development. The lowering of metabolism prolongs development and growth, thus resulting in reduction in mean body size, which in its turn increases the dependence of the animal on the differences between microclimatic and macroclimatic conditions at high altitude. Prolongation of development and growth has also the result of reducing the number of generations. Lowered metabolism is also believed to lower the sensitiveness of these animals to the injurious action of ionizing radiations (3, 5). The high altitude poikilotherms are thus ecologically and physiologically bound to low temperatures and may, therefore, be appropriately described as cryobiont.

The cold stenothermy of altitude poikilotherms is characterized by narrow temperature valance. With increase in altitude, the temperature valance narrows down progressively, so that most species at extreme high elevations are able to tolerate prolonged exposure to low temperatures, but are wholly unable to withstand even a few minutes exposure to temperatures slightly above those normally prevailing in their niches. Unlike the cold-hardy lowland forms, the high altitude forms are characterized by their capacity to utilize the prevailing atmospheric cold, in an effective manner, for counteracting the extreme injurious effects of other environmental factors of high altitude.

Among the poikilotherms, the so called cold-blooded vertebrates represent a transitional group that reacts partly to the atmospheric cold and partly to the low oxygen content of the air, so that fishes, amphibia and most reptiles are excluded at extreme high altitude. These poikilotherms, in which the blood carries the respiratory oxygen, are generally not seriously affected by hypoxia at moderate high altitude.

ii. *Pressure Oriented Animals*

The pressure oriented animals are the homoiotherms that react sharply to the low oxygen content of the air rather than to the atmospheric cold of high altitude. The increasing oxygen deficiency of the air effectively excludes the great majority of the homoiotherms from extreme altitude. Even far below the upper limits of occurrence of the cold-reacting poikilotherms, the homoiotherms exhibit severe pathological symptoms resulting from lack of oxygen. The level of their metabolic rate demands a much higher oxygen requirement than in poikilotherms, but the oxygen supply from the high altitude air is too low for their normal functioning. As the mechanism for regulating and maintaining a fairly constant temperature of the body is based on adequate oxygen supply to the tissues, the high altitude hypoxia is known to induce poikilothermia in some of them (2) (*vide* Chapter XIII, hypothermia in man). Their reactions to the oxygen deficiency fail to bring about an effective adaptation in these animals to the other high altitude factors.

These differences may be traced to the fact that the homoiotherm blood is the carrier of oxygen to the tissue level. In the high altitude homoiotherms we find two principal adjustments to the high altitude environment, *viz.* (i) by increasing the supply of oxygen by functional systems and in the tissue level and (ii) by increasing the resistance of the tissues to the low oxygen of the air. The high altitude mammals are characterized, for example, by high erythrocyte count and reduced size of erythrocyte to facilitate increased oxygen transport through the smaller and more closely packed capillaries than in lowland forms. The flight adaptations of birds may be said to automatically pre-adjust them to both the atmospheric cold and the low oxygen content of the high altitude. It is, therefore, common sight to observe a number of birds, like the cornish chough *Pyrhocorax*, flying at elevations of 6000 m. The snow-partridge *Lerwa nivicola* is also commonly conspicuous at extreme elevations on snow fields on the Himalaya.

Among the specializations of the mammals for life at high altitude the changes in the blood are important. The blood of the llama has, for example, a much larger oxygen-carrying capacity and a correspondingly lower blood viscosity than certain similar animals of the lowland environment (19). This character must be interpreted as a most ideal combination of adjustments that would enable the animal to carry on normal activity under an atmosphere of low oxygen. Some of the other blood peculiarities include the haematocrit and haemoglobin levels, described by Morrison *et al.* (12, 13) in the Chilean and Peruvian Andean rodents, as special adaptation to high altitude life. Chiodi (4) studied the affinity of the haemoglobin to oxygen in some high altitude mammals. The half-saturation point of the oxyhaemoglobin-dissociation in the high altitude *Chinchilla brevicauda* and *Llama paco* corresponds to the carbon dioxide pressure of 24 mm of mercury, compared to the pressure at sea level of 34 mm in rabbits. The myoglobin levels in some tissues of the Peruvian high altitude rodents *Phyllotis darwini*, *Phyllotis rupestris* and *Phyllotis osillae*, at elevations of 3900–4500 m, were investigated by Reynafarje *et al.* (15, 16). The myoglobin concentration in the muscles of the diaphragm and the pooled leg muscles is nearly twice that of *Phyllotis darwini limatus*, which occurs at sea level. The other high altitude species like *Akodon amoenus*, *Akodon jelskii*, *Akodon berlepschii* and *Akodon boliviensis*, studied by the same authors, have shown results similar to *Phyllotis*. The myoglobin content of the Himalayan ibex, that habitually lives at elevations 3000 m, is likewise significantly higher than in the common sheep (Fig. 11–1).

iii. *Some Minor Peculiarities*

The high altitude animals are both terrestrial and aquatic. The specializations in terrestrial forms for life under the extreme conditions are more

143

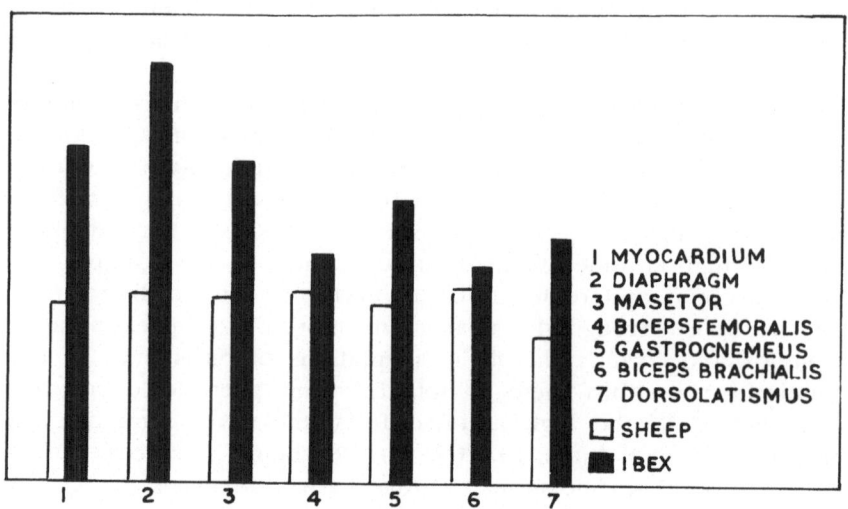

Fig. 11-1. Histograms of the differences in the percentage levels of myoglobin in the common sheep and in Pamir-Himalayan ibex.

marked than the aquatic forms. The typical hypsobiont animals are exclusively geobiont, but the aquatic species at high altitude are essentially cold-adapted ubiquits, which must be classed as hypsophiles rather than as hypsobionts. The absence of marked differences between the highland aquatic communities and the lowland ones is easily explained. The limiting factors for land animals are both cold and low oxygen but for aquatic animals it is only cold. The high altitude geobionts are characterized by the dominance of the endogeous over the epigeous habit, so that the greatest majority of species habitually occur under large stones, boulders and other hypolithic spaces.

In contrast to the animals of the cold boreal latitudes, the high altitude animals are almost without exception characterized by dense body pigmentation. The white and pale colours of boreal animals, interpreted by classical biologists as a protective adaptation of selective value, is in reality the result of the very different environmental conditions prevailing there. The darker pigments of the high altitude animals are the result of the reaction to the intense insolation, the high glare and the action of ultraviolet rays. These factors give rise to the darkening of the exposed parts. The pigments having risen under the action of insolation protect the deeper tissues from further injury by absorption of the radiation. The dark pigments have also an important rôle in absorbing the necessary warmth for metabolism and development from insolation during the hours of sunshine, against the background of white snow.

Certain hypsobiont homoiotherms have dense hairy and scaly covering of body as specializations against cold and aridity. The high altitude mammals of the Andes have remarkably narrow external nasal openings, which insure efficient prewarming of the inhaled air as it enters the lungs. The vision of

144

these animals is also exceptionally keen, as an adaptation for the extensive wide open meadows and snow fields, where unobstructed vision is possible. Typical of these wide areas of high altitude of the Andes are the greatly specialized runners, with slender, long legs and elongated necks, as for example, *Vicugna, Oreailurus, Lynchailurus colocolo, Cavia tschudi, Nothoprocta ornata* and *Nothoprocta pentlandi, Tinamotis pentlandi, Pterocnemia pennata, Thinocorus orbignyianus, Geositta cunicularia, Geositta punensis, Geositta isobellina, Muscisaxicola, Lessonia rufa,* etc. The highly resinous and often also silicious nature of the vegetation of the Peruvian high Andes is reflected in certain peculiarities of the herbivores. These animals have characteristically strong dentition, especially the incisors, for the mastication of the hard food. Large fermentation chambers in the stomach and intestine of *Chinchilla, Abrocoma* and Auchenidae of the Andes, containing symbiotic organisms, subserve also the digestion of cellulose and other food hard to digest.

The diurnal rhythms of the high altitude animals are basically reactions to the atmospheric cold. Forms like *Nycticorax*, which are typically nocturnal in the forest and near sea level, are completely diurnal at high altitude. Groups characterized by two or three generations in lowland become univoltine at high altitude. In some cases a generation requires more than two or three years to be completed.

Though phytophagous species are by no means rare, the hypsobionts at extreme altitudes are predominantly zoophagous. As the elevation increases above the permanent snowline, the abrupt fall in the abundance of green plants is reflected in the increasing dominance of zoophagous animals, until in the extreme elevations of the aeolian zones, above the limits of growth of green plants, zoophagous and scavenging species predominate.

3. Some Typical Hypsobiont Animals

Several groups like Protozoa, Platyhelminthes, Nemathelminthes, Arthropoda, freshwater and land Mollusca, some reptiles, many birds and some mammals have successfully evolved into high altitude forms.

i. *Minor Invertebrates*

The minor invertebrates comprise Protozoa (8), Platyhelminthes (9), Nemathelminthes, rotifers and molluscs (10, 11). With few exceptions, these invertebrates are aquatic or hygrophiles, which are also generally eurytypes. They must be described as hypsophile rather than hypsobiont, but some of the semi-aquatic forms may also be considered as hypsobionts.

The Protozoa *Cyphoderia, Difflugia, Nebela,* etc., common in many high altitude sluggish streams and in lakes, often at great elevations, are typical ubiquits from lowland habitats. The moss-cushions at the edge of glacial streams often harbour Nematodes, Oligochaeta, Ciliata and Flagellata and rotifers like *Philodina, Metopidia, Callidina,* etc., on the Himalaya and the

Alps. The Turbellarians, found at high altitude on the Himalaya, occur almost exclusively in clean, cold melt-water torrents. They are specialized species of widely distributed cold-adapted lowland genera like *Gyrator, Stenostoma, Vortex, Microstoma*, etc. Besides Rhabdocoels, there are also some interesting Triclads in glacial torrents. *Planaria alpina* and *Polyscelis* may sometimes be found in slow-flowing and in stagnant melt-water ponds. These forms are remarkable for occurring in the coldest waters. With perhaps the exception of *Planaria alpina*, the others are typically torren-ticole and occur generally under submerged stones, to the surface of which they adhere by means of a slimy secretion. Though leeches cannot be considered as typical high altitude animals, many species of land leeches are found at elevations of nearly 2700 m on the Himalaya but do not occur higher above. Mollusca are sometimes found on the moss-covered rock-surface, constantly wetted by dripping water, below waterfalls and cascades. The common species belong to *Limnaea, Planorbis, Bythinella, Vitrella, Neritina, Valvata*, etc. *Parvatella, Syama* and *Bensonia* are found at eleva-tions of 2700–3100 m on the Himalaya, but none have been found at higher elevations.

ii. *Arthropoda*

A. Crustacea

The majority of the high altitude crustacea are aquatic and are mostly restricted to glacial lakes on the Himalaya, Alps and the Andes. Without exception, they are ubiquit hypsophiles. The Branchipod *Artemia salina* occurs at elevations of 3600–5000 m in a number of glacial lakes on the Himalaya and on middle Asiatic mountains. *Branchinecta* is abundant in certain melt-water ponds at elevations of 3600–5200 m on Northwest Himalaya and the Pamirs. *Gammarus pulex*, a widely distributed amphipod, occurs in several glacial lakes at elevations of 3700–5335 m on the Himalaya. The glacial torrential streams sometimes contain widely dis-tributed lowland Crustacea like the Cladoceran *Chydorus, Ilyocryptus*, Copepoda *Cyclops* and *Canthocamptus*. *Cyclops fimbriatus*, a cosmopolitan species, is reported to be common in high altitude streams on the Alps. *Canthocamptus* occurs in moss at the stream edge. *Canthocamptus rhaeticus* is a boreo-alpine species from the Alps. Some Ostracoda have also been reported from high altitude waters.

The terrestrial Crustacea *Mesoniscus* and *Protracheoniscus* occur at ele-vations of about 4800 m on the Northwest Himalaya. Other land Isopoda from the Alps occur generally in hypolithic spaces and in soil in alpine meadows. Unlike the aquatic species, these are true hypsobionts.

B. Arachnida

Though the scorpions, as a class, are thermophiles, several species of the temperate and cold-adapted *Buthus, Chaerilus, Scorpiops*, etc. occur on the

146

Himalaya and the Alps. The highest altitude at which a scorpion is known at present from the world is 4300 m on the Northwest Himalaya, where specimens of *Scorpiops rohtangensis* were found under stones, in association with a number of high altitude insects and mites. Being also predominantly a nocturnal animal, the scorpion at high altitude would appear to be at a disadvantage, because all animal activity above ground ceases with sunset. In the hypolithic spaces with nearly uniform microclimatic condition, scorpions find their optima and are rarely if ever seen above ground during the day.

Chelonethida also occur under stones, moss and other organic debris on grassy meadows and in miniature caves, particularly in the vicinity of melt-water. The typical high altitude species belong to *Neobisium*, *Microcreagris*, *Centrochthonus*, *Microbisium*, *Atemnus*, *Tetanotemnus*, etc. The highest altitude at which a pseudoscorpion is known at present is 4300 m on the Northwest Himalaya.

Spiders are without doubt among the most conspicuous members of nearly all high altitude communities. Many species are remarkably at home amidst ice, glacier and snow, on nearly all the important mountain systems in the world. The dominant spiders at high altitude are not, however, the familiar orb-spinners of the forest, but the terricole families like Thomisidae, Gnaphosidae, Drassidae, Clubionidae and Linyphilidae. The high altitude spiders live almost exclusively under large boulders, inside rock crevices and in microcaves. Some semi-xerophile species seem to flourish exceedingly well on barren rock, exposed to the full fury of sun, wind and other inhospitable conditions. The majority of species, including even those that habitually occur in hypolithic spaces, dart regularly on the surface of the snow, during the hours of sunshine, on hunting expeditions. Many like Lycosidae and Salticidae are characteristic of the high aeolian zone of the Himalaya. The altitude record for spider is truly astonishing. Dozens of species are quite common between 3000 and 4000 m. Some unique forms are found only near 6000 m. The highest altitude at which a spider is known is 6700 m on the Himalaya. The common spiders of the Alps include species of *Drassodes* at elevations of 2400–3200 m and *Pardosa* between 2500 and 3200 m. On the Himalaya, *Pardosa* ascends, however, up to 4000 m. *Lycosa* occurs at 6000 m on the Himalaya. *Gnaphosa* occurs up to 3000 m on the Alps and 4000 m on the Himalaya. *Xysticus* and *Thanatus* are reported at 2400–3200 m on the European mountains and have been found up to 4000 m on the Himalaya. *Diplocephalus*, *Macrargus*, *Microneta*, *Micryhantes* and *Epigone* are known at 2500–3100 m on the Alps. *Liocranum*, *Theridium* and others are reported at 3460 m on the Spanish Sierra Nevada, *Heliophanus*, *Aelurillus*, *Euphrys* and *Clubiona* have been recorded at an elevation of 4250 m on Mt. Kilimanjaro.

The mites represent perhaps the dominant arachnids of high altitude and are nearly equally abundant both on land and in water. Some of them flourish at elevations of about 6300 m on the Himalaya and are also reported at 5000 m on the Peruvian Andes. The majority of the high altitude species are geobiont and hygrophiles that occur under stones, in

147

moss, soil or snow surface. Many are predaceous or parasitic. The typical high altitude species belong to *Caeculus, Trombidium, Rhyncholophus, Erythraecus, Rhagidia, Tetranychopsis, Bryobia, Damaeus, Ceratoppia, Hypoaspis, Cyrtolaelapis, Pergamasus, Eugamasus, Bdella, Cyta,* etc.

The Hydracarina like *Halacarus, Neocalonyx, Hygrobates* and *Elyais* are extremely common in most glacial lakes and streams. The first three genera occur at elevations of 5000 m on the Peruvian Andes. *Elyais* occurs on the Himalaya at 4340 m. *Protziella* occurs in springs at 4200 m on the Northwest Himalaya. Several species of *Calonyx* are common in thermal springs, in which the water comes out at 19°C, at an elevation of 3000 m on the Himalaya. *Elyais hamata,* found at an elevation of 4000 m on the Himalaya and at 2400 m on the Alps, is also known to be widely distributed near sea level. In addition to these ubiquits, we have also some stenotypes like *Protziella, Kashmirothyas,* etc., endemic to the Himalaya, which must be described as hypsobionts. Other common Hydracarina include *Thyas, Spodadoporus, Lebertia, Sperchon* and *Feltria.*

C. The myriapod complex

Though many species of Chilopoda, Diplopoda and Symphyla are known at relatively lower elevation on the Alps and other mountains, they are relatively scarce at extreme altitude on the Himalaya. *Lithobius* occurs at elevations of 3500–5200 m on the Himalaya. Some of the important high altitude species belong to *Julus, Ceratosoma, Trimeropheron, Trimerophorella, Polydesmus, Orotrechosoma, Leptoiulus,* etc. *Scutigerella* has been found at 3000 m on the Northwest Himalaya. All the species generally occur under large boulders in moist localities.

D. Insects

A combination of the peculiarities of their body organization, but particularly the respiratory system, enable insects to dominate almost every conceivable habitat and flourish at the highest altitudinal limits of existence of animal life. Their distinctive specializations are in the main directed towards effective utilization of some of the extreme environmental conditions, prevailing at high altitude, so as to counteract the unfavourable effects of the whole complex. The high altitude insects have been fully described in two earlier publications (9, 10), so that only a broad outline of their outstanding features is presented here.

The outstanding peculiarities of the high altitude insects include a remarkable combination of pronounced melanism, reduction and loss of wings and of the powers of flight, reduction in the mean body size, dense body setation, hardening of the integument to retard desiccation, pronounced cold stenothermy, hygrophily, terricoly or obligatory geobiont made of life, prolonged hibernation under snow cover, reduction in the number of annual generations and delayed development and growth, high degree of zoophagous and debris feeding habits, high dependence on snow, etc. Besides black, dark brown or reddish-brown, the other colours commonly met with are red, orange and deeper tones of yellow than in sea-level species. Not only are the ground colours of body and wings darker, but even the spots and markings tend to be larger and deeper in tone at higher altitude than below. With perhaps relatively minor exceptions of some strictly soil inhabiting species, pale coloured insects are unknown at high altitude.

The high altitude melanism is closely correlated with the intense atmospheric cold, high intensity of ultraviolet radiation, glare, intense insolation, crowding of individuals during development and growth and other factors, The heavy body pigmentation is primarily the result of the action of ultraviolet radiation. Having thus risen, it serves to protect the delicate tissues underneath from further injury by cutting off the high intensity radiation. The pigmentation serves also to absorb the necessary warmth from the sunshine. It is common experience to find swarms of insects on warm rock surface and on snow fields during the hours of sunshine, with the body oriented in such a way as to expose the maximum surface to sunrays.

In nearly all lowland ecosystems, winged insects as a rule outdominate the wingless or the short-winged forms. The vast majority of the high altitude species have vestigial wings or are fully apterous. Even the small numbers of species, in which wings are well developed, rarely ever fly. Flightlessness and the wingless conditions are related to the prevailing atmospheric cold, the high wind velocities, endogeous habits, delayed metamorphosis and other conditions. In the Northwest Himalaya almost 50% of the hypsobiont species are flightless at elevations between 3000 and 4000 m. The apterous condition is met with in 60% of the species at elevations above 4000 m. Near the permanent snowline and at the upper limits of insects life, nearly all the species are wingless. Almost 85% of the Pterygota on Mt. Kilimanjaro are either brachypterous and wholly unable to fly or are apterous (10, 11).

While large and medium-sized insects are by no means completely absent, giant species so characteristic of the humid tropical forests, are wholly absent at high altitude. There is on all mountains a very pronounced tendency for reduction in the mean body size with increase in altitude. The size reduction is evidently related to the delayed development, intense environmental cold, loss of wings, crowding in microclimatic refugial niches, etc.

The high altitude insects are typically cryobionts, characterized by their capacity not merely to survive, but also to continue their normal development at temperatures at which most lowland species pass into cold stupor. Low temperature reduces the risk of desiccation. Nearly all the species are concentrated at the snow edge, as the snow is the only source of moisture for them. The snow serves also as an effective protective covering during hibernation in winter. The insects use the snow fields as hunting and feeding ground. Insect life at high altitude may be said to be possible only because of the cold and snow.

The high altitude insects are pronounced geophiles, geobiont and endogeous, so that even the few planticole species tend to remain close to the ground. All activity above ground is severely restricted to the hours of bright sunshine. Crepuscular and nocturnal species are wholly unknown. The hibernation of insects under the protective cover of winter snow ends only when the snow starts melting with the advent of summer. The above ground activity is initiated only after the ground has been sufficiently warmed by summer insolation. On bright sunny days, with little or no wind, the surface activity begins an hour or two after sunrise. As the morning advances and with further rise of the air temperature immediately above ground, activity diminishes, until there is a lull in the early afternoon. There is often a partial resumption of activity in some insects in the early evening, followed by cessation of all above-ground activity with sunset.

At extreme high altitude, only a small proportion as species is phytophagous. On the Northwest Himalaya, for example, no more than 3% of the high altitude species are phytophagous above an elevation of 5000 m. The bulk of the species either feed on organic debris or carrion or are predators.

The dominant high altitude insects are Plecoptera, Coleoptera, Diptera. Lepidoptera and Collembola, but other familiar orders like Ephemerida, Orthoptera, Dermaptera, Heteroptera, Hymenoptera, Trichoptera and Thysanura are not uncommon.

(a) Ephemerida

The Ephemerida larvae are almost exclusively torrenticole and cling to submerged stones in glacial torrents; only extremely rarely are they found in glacial lakes. The typical high altitude species belong to *Ameletus, Rithrogena, Ecdyura, Ecdyonurus, Caenis, Baetis, Cloeon, Palingenia, Ephemera* and *Iron*. The subimago and adults may often be found scattered in enormous numbers dead on snow fields during summer. Their dead bodies serve as food for a variety of carrion feeders.

(b) Plecoptera

The Plecoptera larvae breed in cold glacial torrents even up to the permanent snowline, at temperatures between 0.5 and 4.0 °C. The adults usually emerge during the coldest part of the day. The fully mature naiads crawl out of the water late in the evening, when the water level in the stream is low (because little snow melts then in absence of sunshine), climb on to stones and rock surface nearby and emerge as adults a little later, leaving behind the empty casts. While some species are phytophagous in their larval life, most others are predaceous on the larvae of Ephemerida, Chironomidae, etc. Nearly all species require more than one year, often two to three years, to complete a generation. After mating, the females crawl back to the edge of the stream to deposit eggs. The dead and dying stoneflies on rock and snow are devoured by swarms of beetles, birds, etc. The common high altitude species belong to *Capnia, Chloroperla, Nemura, Perlodes, Amphinemura, Rhabdiopteryx* and *Kyphopteryx*. Species of *Rhabdiopteryx* have been found breeding at an elevation of 5000 m. Some of the species are typically brachypterous or apterous at extreme elevations. They then become semi-terrestrial and lack also ocelli. Subapterous and apterous *Capnia* are common above 5000 m on the Himalaya.

(c) Orthoptera

Though some remarkable Tettigonids are known at high altitude, the majority are Acridids. The highest altitude record for the order is 4900 m on the Himalaya, but the nymphs of *Dasynema* were found at an elevation of 5490 m. *Acrodectes philophygus* is reported at 4420 m on Mt. Whitney. Most high altitude grasshoppers are flightless or have more or less reduced wings or are apterous. *Podisma*, *Bryodema* and *Conophyma* are, for example, apterous on the Himalaya. The common high altitude species belong to *Bryodema*, *Podisma*, *Aeropus*, *Gomphocerus*, *Aeropedellus*, *Xanthippus*, *Melanopulus*, *Dosciostaurus*, *Dicranophyma*, *Conophyma*, *Spingonotus*, etc.

(d) Coleoptera

The Coleoptera belong to Carabidae, Staphylinidae, Tenebrionidae, Dytiscidae, Hydrophilidae, Histeridae, some Chrysomelidae and Curculionoidea. Coccinellidae and many Scarabaeoidea are also often found. The species are geophiles or geobionts and habitually occur under large stones, in cavities under deeply sunk boulders, in rock crevices and in soil, almost exclusively in the vicinity of glacial streams and the snow edge. Debris feeders, carrion feeders and predators predominate over phytophagous species. Most species occur at elevation between 2500 and 4000 m. The highest altitude record for the order is 5600 m, at which elevation occurs the Staphylinid *Atheta* (*Dimetrota*) *hutchinsoni* on the Himalaya. Many Carabidae like *Carabus*, *Nebria*, *Cychrus*, *Amara*, *Trechus*, etc. are also typical of the high altitude ecosystem and of the high boreal latitude. Many of them have also circumpolar distribution. The tribe Anilini, which are amongst the smallest forms and often measure no longer than 1–2 mm, are completely flightless and are among the most conspicuous members of the high altitude Carabids. Nearly all the species are predaceous and even among the generally debris-feeding Staphylinidae, the typical high altitude species are predatory. The Tenebrionidae seem particularly to be suited for life at high altitude. Some species of *Blaps* occur at 5000 m and *Ascelosodis* occurs at 4400–5600 m. The typical high altitude tenebrionids are *Itagonia*, *Prosodes*, *Platyscelis*, *Bioramix*, *Myatis*, *Laena*, *Scyathis*, *Opatrum*, *Cyphogenia* and *Gnaptorina*. Some remarkable endemic Chrysomelidae of the Himalaya include *Apaksha* and *Chaetocnema alticola* found at 4575 m. The hypsobiont Cucrulionids belong mostly to genera like *Otiorrhynchus*, *Blosyrodes*, *Catapionus*, *Pissodes* and *Scepticus*. *Macropscoelorum* is reported at an elevation of 4976 m on the Andes. This is at present the highest record for the family. Most of the high altitude Curculionids of the Himalaya and of the Andes are brachypterous or apterous. Almost without exception they occur under stones on

151

grassy meadows. Some interesting Scarabaeidae like *Copris, Aphodius* and *Onthophagus* are often found actively rolling dung, on sunny slopes up to almost 4000 m on the Himalaya and the Pamirs. The genus *Taraoceraster* is peculiar to the high altitude of the Andes.

(e) Lepidoptera

The high altitude Lepidoptera are the Papilionidae, Parnassiidae, Nymphalidae, Satyridae, Pieridae, Lycaenidae, Arctiidae, Geometridae and Noctuoidea. Though most species are confined to elevations of 2500–4000 m, a number of interesting species breed at 6000 m. Many species are indeed entirely restricted to extreme elevations between 5000 and 6000 m and never occur below 4000 m. All species, including also the otherwise nocturnal Geometridae and Noctuoidea, are strictly diurnal at high altitude. Though the caterpillars of some forms like *Scoparia* and *Callimorpha* breed on Lichens, they are mostly grass breeders. In nearly all cases the caterpillars remain under stones or inside rock crevices and generally never appear above ground on the surface of their food plants as in the case of the species on the plains. The most common high altitude Lepidoptera are *Parnassius, Argynnis, Erebia, Pieris, Colias, Gnophos, Larentia* and *Dasydia*.

(f) Diptera

The Diptera are not only among the dominant insects, but also occur almost upto the highest limits of insects life on the Himalaya; they are, for example, abundant at 6000–6300 m. The majority of species breed in organic debris at the snow edge or close to melt-water streams and glacial lakes. A large number of species at extreme high altitude are carnivorous and active predators. The dominant families at these elevations are Stratiomyiidae, Empididae, Ragionidae, Asilidae, Dolichopodidae, Calliphoridae, Anthomyiidae and Tachinidae. Among the flower-visiting insects at high altitude, the Diptera stand foremost. Not only Syrphidae, but the adults of most other families regularly visit flowers. The Diptera become active after hibernation much before most other groups.

Many species of the high altitude Tipulidae breeding at 5200 m are subapterous or apterous. The glacial torrential streams are the special habitat of the larvae of Blepharoceridae, Deuterophlebiidae and Simuliidae. *Stratiomyia* and *Odontomyia* occur at elevations of 3500 m. The Tabanids *Sziladnus* and *Tylostypica* are found at 3400–3700 m, on the Pamirs. The Thereviid *Reinigellum* is known at 4400 m on the Pamirs. The Syrphids include *Melanostoma, Platychirus, Eriozona, Mallota, Temnostoma, Criorrhina, Chrysotoxum, Arctophia, Syrphus, Heliophilus* and *Volucella*. Muscoid flies are extremely abundant and include *Chortophila, Limnohora, Hydrophorina, Anthomyia, Phaonia, Trichopticus, Hydrota, Bithoracochaeta, Pollenia*, etc.

(g) Other orders

Some peculiar wingless Dermaptera like *Anechura* and *Forficula* occur up to 4300–4900 m. *Grylloblatta* is found up to 3000 m on the mountains of Japan and North America. Thysanura are often abundant on barren rock, inside rock crevices and under stones. Small swarms of *Machilinus* and *Ctenolepisma* may be found feeding on lichen and other vegetable debris at 3600–5800 m. Collembola are *par excellence* the high altitude insect that swarms on snow field and often imparts to it a black colour, which is conspicuous even from a distance. Collembola occur at 6300 m under stones, among roots, in soil, moss cushions, on snow and surface of glacial lakes. The common forms belong to *Hypogastrura, Onychiurus, Isotoma, Proisotoma, Orchesella, Lepidocyrtus, Bourletiella, Cyphoderus, Entomobrya,* etc.

iii. *Vertebrates*

The vertebrates as a whole and poikilotherms in particular are extremely poorly represented at high altitude. The Cyprinid fish *Diptychus maculatus* occurs in the R. Indus and part of the source streams of the R. Chenab in the Northwest Himalaya, in waters with temperatures as high as 12 °C, at elevations of 3300–4000 m. Unconfirmed Chinese reports of certain Tibetan loaches in mountain brooks at elevations of 5200 m have appeared in *Science News* in 1977. Some of these elevated regions, where fish are reported, have relatively hot summer and relatively high water temperatures. Although some species of *Bufo* are known to occur at high altitude of the Puna Andes, the high altitude environment excludes the Amphibia in other parts of the world. The highest altitude of record of Amphibia is 4572 m in Mexico, from where the salamander *Pseudoerycea gadovii* is recorded. The extreme conditions of the high altitude environment exclude likewise the Chelonia at elevations above 2000 m. The lizards and snakes, several species of which are found on the Himalaya, represent essentially recent secondary derivatives of the ancestral stocks, which differentiated in the arid steppes of Middle Asia at much lower elevations. At high altitude on the Himalaya, they are typically petricole during the daytime and hypolithic at night. The large and warm nunatak rocks at high altitude, exposed to the sun, attract large numbers of lizards and snakes for basking in the sun and for absorbing the heat that is being radiated from the rock surface. Some of these Himalayan lizards like *Leiolophisma ladakense* and *Leilophisma himalayanum*, often occur from above the timberline up to 5500 m. *Sibynophis collaris* occurs up to 3000 m. The snake *Ancistrodon himalayanum* is reported at 3000 m in the Eastern Himalaya and has been found even as high as 4800 m on the Northwest Himalaya, representing perhaps the highest altitude record for snake. *Crotalus triseriatus* occurs on barren ground at an elevation of 4528 m on Citlalepetl (18). The lizard *Sceloporus microlepidotus* is reported as actively pursuing wind-blown insects at elevations of 4816 m on the same mountain. Pearson (14) has described the habits of the lizard *Liolaemus multiformis multiformis* at an altitude of 4900 m on the southern Peruvian Andes.

Many birds which occur exclusively at high altitude exhibit, however, peculiarities not necessarily specific for the high altitude environment, but for other secondary and local factors, in the Andes or parts of the Himalaya. The wide-open elevated regions of South America are inhabited, for example, by running birds like *Vicugna*, *Oreaulurus*, *Lynchailurus*, *Cavia*, *Nothoprocta*, etc. Characteristic of the snow-covered areas of the high Himalaya are birds like the snow partridge *Lerwa nivicola*, *Pyrhocorax* which may be observed flying above 6300 m. Various species of Himalayan eagles occur a little lower.

Among the mammals, though some forest-zone species like *Chimarrogale himalaica* may be found up to 4500 m on the Himalaya, the typical hypsobiont forms descend only exceptionally to within the upper reaches of the forest. As with most other animals, the mammals are also petricole, hypolithic and troglophile at high altitude. This would explain perhaps the general dominance of rodents among the high altitude mammals.

Most high altitude mammals hibernate in winter, but some move down to lower valleys. The rodent Akodon hibernates on the Andes, but *Hippocamellus* comes down to lower elevations during winter. Like in other high altitude groups, the mammals of extreme high altitude are generally predators. The typical carnivores of the Himalaya include the snow-leopard *Unica unica*, *Lynx lynx* and *Otocolobus manul*. The common mammals found above the timberline on the Himalaya include *Nectogale*, *Otonycteris*, *Eupetaurus*, *Arctomys*, *Sminthus*, *Cricetus*, *Lagomys*, *Pantholopus*, *Ovis*, *Capra* and *Moschus*. On the inner ranges of the Himalaya and partly extending into western Tibet and the Pamirs may be found *Vulpes pusilla*, *Martes foina*, *Putorius putorius larvatus*, *Ursus arctos*, *Ursus torquatus*, *Arctomys himalayanus*, *Arctomys hodgsoni*, *Arctomys caudatus*, etc. *Ovis hodgsoni*, *Ovis poli*, *Ovis vignei*, *Ovis nathura*, *Capra sibirica*, *Capra falconeri* (the largest wild goat of the Pamir), *Capra aegagrus*, *Nemohaedus bubalinus*, *Cemas goral* and *Moschus moschiferus* are among the exclusively high altitude Himalayan mammals. The wild yak *Bos grunniens* is a short-legged, blackish-brown, bison-like form, with drooping head, hair hanging in shaggy fringes on the flank, shoulder and thigh. Its dense, soft and closely matted underfur affords additional insulation for the body warmth. It roams the desolate regions of the inner Himalaya and Tibet, at elevations of 3300–4500 m. The llama amd the alpaca are the other well known high altitude mammals from South America.

4. Lowland Animals at High Altitude

Two groups of lowland animals are frequently met with in the highland ecosystem, *viz*, the aeolian derelicts and the summit haunting forms.

The aeolian derelicts comprise the wind-blown organic particles on mountain slopes. The presence of such air-borne material on high mountains has been known since 1881 (9). In these derelicts we find spores of fungi and of Protozoa, pollen grains, seeds and small fruits of various plants, bits of dry leaves and bark, mites, spiders, aphids, jassids, membracids, thrips, aleurodids, males of coccids, dragonflies, diurnal beetles, adult cutworm and other small moths, many species of butterflies, a host of Diptera, mantids, grasshoppers, winged ants, besides fragments of the wings, legs and other disintegrated parts of all sorts of insects, encysted ova of Nematodes and other miscellaneous chitinous fragments. Sometimes even large-sized and heavy-bodied wasps, Cantharids, Cerambycids, locusts, etc. have been found in the wind-blown material on the Himalaya and North America mountains. The spiders thus found include not only the light-bodied ballooning orb-spinners from bushes in the plains, but even the much heavier ground-nesting thermophile species from the hot desert areas.

The aeolian derelicts occur at elevations from the treeline to the highest summits. Bacteria, fungal spores, yeast and soil nematodes have, for example, been reported at elevations above 8000 m on Mt. Everest. The presence of wind-blown particles is, however, particularly marked on the Himalaya at elevations of 5000–6000 m and higher, on the southern slopes. It seems possible that some of the material found on the high peaks have dropped from higher layers of the air, by slow gravitational settling down. The upper altitudinal limits to which air currents can lift up particles from the plains can only be conjectured. Part of the aeolian deposit comes no doubt from the warmer lower and outer valleys, but the major portion is brought from the far-off plains. The load of inorganic and organic matter contained in the air during summer is truly enormous. In a short period of about twenty minutes, over four hundred dead insects of the plains were blown in on an area of snow field, about 10 m² at an elevation of 3500–4000 m during May–June on the Himalaya. A small fraction alone of the total air-borne material must actually be deposited on the Himalayan snow fields, the bulk being scattered randomly over much lower elevations.

Strong and violent winds no doubt account for the great bulk of the air-borne load, but not necessarily all of it. Even when strong surface winds are absent in the plains, the warm updraft air currents, rising from the heated ground during summer, are strong enough to pick up and carry all sorts of light seeds, ballooning spiders, flying insects and other forms from the ground also. The spiders and insects thus lifted up become chilled on reaching the upper layers of air and rendered incapable of navigating themselves. They are thence passively carried about, till some of them fall on the Himalayan mountain sides, dead and completely frozen.

Although the deposition of the wind-blown matter is about equally frequent on snow fields and on the barren rocky ground, the deposit on snow is particularly rich. On the surface of snow the particles are securely trapped by adhesion and adsorption to the snow granules. The deposit is also often covered by fresh snowfall. In course of time, the particles come to be securely entombed inside the snow and glacier ice. The accumulations of the refrigerated and well preserved wind-blown organic derelicts over a period of several years, often even centuries, may be found in stratified layers inside the glacier, especially at the snout or exposed sides and in crevasses. The best known of such accumulations is perhaps the grasshopper glacier of Montana in North America (6). This glacier contains accumulations, at least six hundred years old, in stratified layers, of the prairie grasshopper *Melanoplus*.

ii. *Summit-Seeking Species*

The summit-seeking species exhibit an enigmatic behaviour of active movement from lower valleys and distant plains to the high mountain massifs. The aeolian derelicts reach the highland areas more or less dead and frozen, but the summit-seeking species arrive alive and may even show considerable activity on the mountain. Some of them actively fly about, dance in the air or also crawl about on the ground, apparently aimlessly for a time. On rare occasions they have even been seen in coitus. Eventually most of them become, however, sluggish and pass into dormancy. While the aeolian derelicts are composed of a miscellaneous assortment of plant and animal, the summit seeking species are exclusively winged and actively flying insects of the plains. The movement of these insects from lower elevations to the mountain summits is evidently a voluntary action. Though favourable winds and air currents may assist them, they are not completely dependent on such air currents for transport to the mountain. The ascent of the mountain is undertaken even when the air is calm and there are apparently no warm updraft currents. The summit seeking activity begins only after summer and is maximum in autumn. The vast majority of the summit-seeking species appear to prefer moderately high altitudes, upto 5000 m and only exceptionally do we find some of them ascending above the permanent snowline to elevations of 6000 m on the Himalaya.

The habit of summit seeking has been observed in many different groups like winged ants, beetles and flies. A bewildering variety of explanations have been put forward by different workers regarding this behaviour, but it must be admitted that we are still largely ignorant about the basic factors underlying this urge and the precise significance of the summit haunting (10).

At certain times of the year, enormous swarms of the adults of winged ants, moths, ladybird beetles and flies rise from the outer and lower valleys of the Himalaya, from the forest and from distant plains of India, and move steadily to the mountain summits. In these swarms the insects apparently seem to show a more or less decided and as yet unexplained preference for some mountain peak or ridge over numerous others, which may often be quite closeby and be apparently equally or even more readily accessible and exposed. None of the species that swarm on mountain tops is resident on the mountain, but only regular visitors. Swarms of Coccinellid beetles, especially

Coccinella, Adonia, Hippodamia and some other genera, move upward to high mountains from the forest below and even from far-off valleys and plains. The beetles cover considerable distance and ascend to elevations of nearly 4500 m on the Himalaya. The beetles condense in great mass assemblages of 5000–10000 individuals under some suitable sheltered stone, boulder, overhanging rock or snowfield. The mass assemblages are covered by the winter snow, often several metres thick, and the beetles undergo hibernation during winter under the snow. With the coming of summer and the melting of the winter snow, most of the dormant beetles revive and move back to lower valleys and plains for breeding. At this time hundreds of pairs may be seen in coitus, crawling all over and taking to wings. Large numbers of individuals naturally perish and the dead beetles in various stages of disintegration generally scatter the snowfield and are often carried down in melt-water streams. The swarms seem habitually to assemble in the same area year after year. Accumulations of dead and frozen beetles from the swarms of previous years occur along with the newer and live individuals of current swarms. Perhaps the largest and the highest coccinellid mass assemblage in the world is the one in which over two million individuals of *Coccinella septempunctata* and *Hippodamia*, found by me at an elevation of 4260 m on the Himalaya (9).

iii. *Ecological Significance*

The ecological significance of the presence of the lowland animals at high altitude is great. The dead bodies of these animals provide an abundance of food to a variety of the high altitude organisms. The accumulated aeolian derelicts and the ladybird beetle mass assemblages are devoured by necrophagous Collembola, Staphylindae, Tenebrionidae, Diptera, Acarina, birds, bears, rodents, etc. It is reported that the grizzly bear feeds on the armyworm moths *Agrotis* and on the coccinellid masses on mountain summits on Montana in North America. The aeolian derelicts and the summit seeking mass assemblages constitute the most important source of organic nitrogen in the melt water for the high altitude plants.

It seems remarkable that both at extreme high altitude on mountains and in the great abyssal darkness of the sea, animals should depend on the sea-level life for their very existence. The ultimate basis of all life on the earth at present lies in the photosynthesis of carbohydrates by green plants, for which three things are absolutely needed, *viz.* water, carbon dioxide and sunlight. As the availability of these things diminishes, the abundance of green plants also falls. An abundance of green plants is associated with an equal abundance and diversity of animals, which together constitute a self-regulating autotrophic system of life in the plains and proximity to sea level. As the green plants become scarce and disappear eventually both in the abyssal depths of the sea due to total absence of sunlight and at extreme high altitude on mountains due to excessive cold, aridity and deficiency of carbon dioxide, the foundations of the system shift from the autotrophic to the heterotrophic (Fig. 11-2). While in the autotropic lowland systems

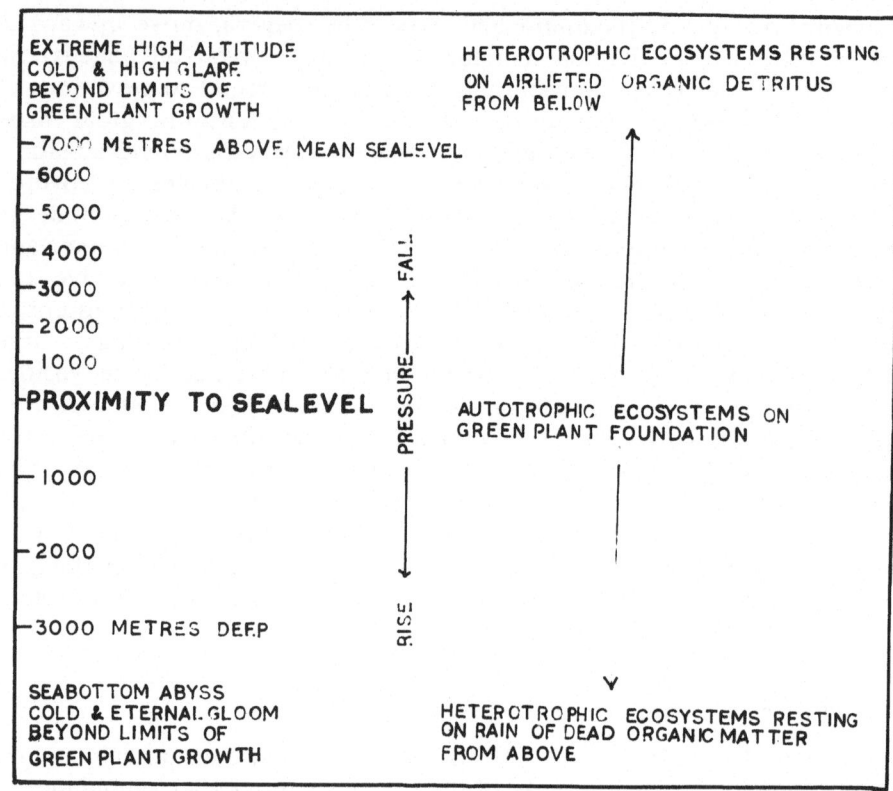

Fig. 11-2. The absence of green plants at extreme high altitudes and in the bottom of the deep sea, animals depend on organic detritus from near sea level. Low pressure and high pressure environments lie on the borders of life on Earth.

zoophagy and scavenging are merely one of the many nutritional interrelations, they come to be the only nutritional interrelation in the abyss and at high altitude. The abyssal animals, in the eternal darkness and cold of the deep sea bottom, depend for their food literally on the 'rain' of organic food particles derived from the death of animals of the surface of the sea or they devour each other. The high altitude animals, in their world of high glare, aridity and cold on the tops of high mountains, likewise devour each other or depend literally on the food particles air-lifted from near sea level. Life on the outskirts of the biosphere leans heavily on the centre from where it radiated both below and above.

5. References

1. Altmann, P. L. & D. S. Ditmar. 1961. Blood and other body fluids. *Fed. Amer. Soc. expt. Biol. Washington*, 1–191.
2. Bhatia, B., S. George & T. L. Rao, 1970. Hypoxic poikilothermia in rats. *J. appl. Physiol.*, 27 (5): 583–586.
3. Bychkovskaia, I. B. & G. K. Ochinskaia. 1960. Protective effect of hypoxia at various radiation doses. *Biofizika*, 5 (4): 468–473 (pp. in English translation 532–538.).

4. Chiodi, H. 1962. Oxygen affinity of the haemoglobin of high altitude mammals. *Acta Physiol Latinoamer.* 12 (2): 208–209.
5. Duggar, B. M. 1936. Biological effects of Radiation. New York: McGraw-Hill Book Company Inc. Vol. 2.
6. Gurney, A. B. 1952. Grasshopper glacier of Montana and its relation to long-distance flight of grasshopper. *Ann. Rep. Smithsonian Inst.*, pp. 305–325.
7. Lamiger, H. 1971. Über das Vorkommen von Schalenamöben (Rhizopoda testacea) in alpinen Polster- und Rosettenpflanzen. *Zool. Anz.*, 186 (5/6): 332–335.
8. Lamiger, H. 1972. Ein Beitrag zur Kenntnis der Hochgebirgstestaceen Österreichs–*Ark Protistenk.*, 114: 101–151.
9. Mani, M. S. 1962. Introduction of High Altitude Entomology, London: Methuen & Co.
10. Mani, M. S. 1968. Ecology and Biogeography of High Altitude Insects. The Hague: Dr W. Junk – Publishers.
11. Mani, M. S. 1974. Fundamentals of High Altitude Biology. New Delhi: Oxford & IBH Publishers pp. 196.
12. Morrison, P. R., K. Kerst & M. Rosemann. 1963. Hematocrit and hemoglobin levels in some Chilean rodents from high and low altitudes. *Intern. J. Biometeor.* 7 (1): 45–50.
13. Morrison, P. R., K. Kerst, C. Reyna Farje & J. Ramos. 1963. Hematrocrit and hemoglobin levels in some Peruvian rodents from high and low altitudes. *Intern. J. Biometeor.*, 7 (1): 51–58.
14. Pearson, O. P. 1954. Habits of the lizard *Liolaemus multiformis* at high altitudes in southern Peru. *Copeia*, 2: 111–116.
15. Reynafarje, B. 1962. Myoglobin Content and enzymatic activity of muscle and altitude adaptation. *J. appl. physiol.*, 17 (1): 301–305.
16. Reynafarje, B. & P. Morrison. 1962. Myoglobin levels in some tissues from wild Peruvian rodents native to high altitude. *J. Biol. Chem.*, 237 (9): 2861–2864.
17. Schaller, G. B. 1977. Mountain monarchs: Wild Sheep and goats of the Himalaya. Chicago Univ. Press XVIII 426.
18. Swan, L. W. 1967. Alpine and aeolian regions of the world. In Wright, H. E. Jr. & W. H. Osburn, Arctic and Alpine Environments. Indiana Univ. Press 3: 29–54.
19. Whittembury, J., R. Lozano, C. Torres & C. Monge. 1968. Blood viscosity in high altitude Polycythemia. *Acta Physiol. Latinoamer.*, 18 (4): 355–359.
20. Wiesen, W. 1973. Effects of temperature on Ectothermic organisms. Berlin Heidelberg/New York: Springer Verl. pp. 298.

Man in highlands ecosystem: physiology of native highlanders

M. S. Mani & A. Pardo

1. Introduction

Man at high altitude has most appropriately been described as *eine habitat-fremde Art* or ecological anomaly (11). Human anatomy and physiology, which are products of arboreal ancestry and subsequent evolution in a forest environment, are in complete disharmony with the extreme conditions of semi-attenuated air, prevailing in the treeless world of high elevations. The high altitude is an unnatural and indeed a hostile environment for man. This has not, however, prevented man from venturing extensively into the highland ecosystems.

Nearly ten million peoples are permanent residents of high altitude regions, at elevations of 3600–5300 m, in the world. Almost 80% of these peoples live on the Peruvian and other parts of the Andes in South America. The remaining are found on the Himalaya, Tibet, Ladakh, Pamirs and the Abyssinian Highlands. About 5% of the population of Switzerland live at elevations of 1000 m and only 0.5% at elevations of 1500 m. The capital city of Yemen in Arabia is situated at an elevation of 2130 m. Most of Abyssinia lies at elevations of 2000 m and Addis Ababa is at 2400 and Ankaber at 2600 m. The greatest altitude, at which human beings live permanently at present, is 5334 m in the mining settlement at Aconquilcha on the Andes. The Incas were also high altitude inhabitants of the Peruvian Andes.

The Incas' empire extended over large areas of the Andes. The Incas' Capital Cuzo was at an elevation of 3400 m. Fortifications were erected at an elevation of 2440 m at Majchu and there are also ruins of forts and temples of the Incas at Sacsayhuaman at an elevation of 3580 m.

Economic life in Peru and Bolivia is concentrated at high altitude areas since centuries. Mining industry flourishes, for example, at Cerro de Pasco (4330 m), Potasi (4070 m) and Oruzo (3760 m) in South America. The miners from Cerro de Pasco regularly ascend up to 5790 m every day for work. In Ladakh, gold is mined at Tok Dschalung at 4880 m at the headwaters of the R. Indus. In Tibet and Ladakh ancient and great Buddhist monasteries have flourished for centuries at elevation of 3600–4200 m. In Nepal the Sherpa and other people are permanent residents of high altitude and cultivate paddy in stone-terraced fields. Indian shepherds like the *gaddis* camp regularly in the open, with their sheep, during summer on the Himalaya, at elevations of 4200–5500 m and often even higher and may be considered in a sense as semi-permanent high altitude residents. Besides these permanent residents, Hindu ascetics, who have developed the capacity of voluntarily controlling some of the involuntary functions of the human body by practising *hata yoga*, have the remarkable power of living, from

161

time immemorial, without any special protective clothing and without special food, in the open under near zero air temperatures, at altitudes between 4000 and 5500 m on the Himalaya, without apparent harm. Hundreds of pilgrims, men and women of middle and old age, from the hot and dusty plains of India, regularly ascend to various holy shrines like Badrinath, Kedarnath, Amarnath, etc. on the Himalaya, which are often situated at elevation of 3300–4000 m. Military operations have also been undertaken by the Indian army at elevations of 4000 m. The Indian Border Roads Organization has recently constructed the second highest mountain motorable road in the world, crossing the Himalaya at an elevation of about 4890 m. The Indian Air Force regularly operates the highest airfield in the world at almost 4000 m on the inner Himalaya, besides several other air stations at about 3000 m. Mountaineers of various countries have of course successfully reached elevations above 8000 m on the Himalaya and Karakoram. Mountaineers are highly motivated people, who deliberately expose themselves to the upper limits of endurance in hostile environment for relatively short periods. Supersonic air pilots (modern commercial aircraft is pressurized and is provided with emergency oxygen supply) have no doubt climbed in unpressurized aeroplanes to much higher altitudes. Balloonists have frequently ascended to nearly 8000–12,000 m.

Early experience showed that flight to 5000 m in balloons or open planes produced symptoms of mild oxygen deficiency, even without exertion. At 8000 m, nobody could escape disturbances, even though individuals vary greatly in their tolerance to altitude. The U.S. Airforce now requires oxygen to be used on flights above 3050 m. It considers 5000 m to be in the zone of 'disturbance', and 6000 to 7000 m to be the 'critical zone'. In unpressurized aircraft, using pure oxygen and tight fitting masks, modern pilots cannot ascend above 12,000 m. Above this some form of pressurized breathing equipment has to be used. Experimental evidence shows that this is due to a sharp decrease in arterial oxygen saturation, that is to say, a lessened oxygen content of the blood. Above 19,000 m, called the Armstrong line, dysbarism or decompression affects pilots. At about this altitude, body fluids begin to boil. Pilots must use sealed pressurized cabins if they are to survive exposure to such altitudes. The problems of such high altitude aeroplane and rocket flights belong rather to the realms of aviation medicine, aerospace and exobiology than to high altitude biology and are outside our scope.

There are thus two principal classes of human populations in highlands viz. the native inhabitants who are permanent residents of the highlands regions and the short-term visitors from lowlands viz. mountaineers and explorers. In this chapter we deal with the highland natives and their peculiarities.

2. Natives of Highlands

Broadly speaking, there are two major highland regions in the world, where large native human communities have long existed viz. Tibet and the South

Plate 12-1. Native highlanders: 1. Tibetan nun. 2. Himalayan farmer.

American Andes. It is interesting to note that in both these areas the population is derived from the Mongoloid stock. The South Americans are descended from remote Mongoloid ancestors, who emigrated from Asia via the Bering strait to America during perhaps the late Pleistocene times. It is, therefore, not surprising that the general physique of modern highland natives of Tibet and the Andes is determined largely genetically. It is at the same time also evident that long residence of thousands of years has brought about anatomical and physiological adjustments to the extreme conditions in these peoples. The literature on the anatomical and physiological peculiarities of the native inhabitants of diverse highland regions is extensive and greatly scattered. Although some attention has been paid to the highland natives of Tibet and Nepal (5, 10, 19), much of our knowledge of the socalled high altitude adaptations is based on investigations in the Peruvian and Bolivian Andes of South America (1, 2, 3, 4, 7, 8, 9, 12, 13, 14, 15, 16, 18, 20, 21, 22, 23, 24).

The mongoloid ethnic basis of some of the peculiarities of the Andean natives of South America is demonstrated by the predominance of the blood group O in all populations. The natives of the northern Peruvian Andes belong wholly to this group. Nearly 95% of the Bolivian Indians belong to

163

Plate 12-2. Native children of highlands. 1. 1. Tibetan children. 2. Bolivian children.

the group. Similarly 85% of the native inhabitants of the Quechua Indians of Milpo and Coloquifirca in the Peruvian Andes, permanently resident at elevations of 4000 m, have the O-group. In the Andean Altiplano, 93% of the Aymara Indians belong to the same blood group and 100% the Lamistas tribe of the Peruvian Andes belong to the O-group. These facts would appear to suggest that at least some of the characters of the general

164

Plate 12-3. 1. Altiplano, area outside La Paz in Bolivia 2. Mountain farming by native highlanders of one of the high Himalayan 'valleys' at elevation of 3600 m.

physique, attributed at present to the effects of high altitude environment, may perhaps be based on the ethnic factor. At the same time, there seems to be little doubt that the ethnic characters are superimposed by others, which have developed as a reaction to permanent residence at high altitude. It would be safe to conclude that since the Tibetans and Andean native highlanders are both of the Mongoloid stock, their physique is a composite of ethnic equipment and of partial (incomplete) adaptation (especially their socalled barrel chest and short stature).

The mechanisms of adaptation and the alterations caused by their prolonged life at high altitude, affect almost all aspects of body function,

165

although some of these points have not yet been studied carefully. Systems that have been largely studied refer only to mechanism of respiration and cardiovascular, hemopietic mechanism and some endocrine glands.

3. General Physique

The general physique, characterized by short stature, is perhaps determined by partly ethnic and partly high altitude factors. The generally short stature of high altitude inhabitants, nearly all over the world, seems to be correlated with the slow prenatal and postnatal growth. The relatively small weight at birth in these natives is essentially a response to the lower oxygen content of the air at high altitude. Exposure to hypoxia from infancy causes the organism to undergo adaptations. It would perhaps be better to say that the organism benefits from certain changes which make it to stand the rigors of highland life better than before. Studies at over 2000 m have shown that as a rule children are slightly smaller than at sea level. The circumference of the thorax is normally greater, which shows that the increase in breathing capacity comes quite early, although the rhythm is analagous to that of sea level inhabitants. The weight of the children tends to be the same as at sea level, although some investigators have found that weight is somewhat less. Highland children grow more slowly than sea level children, resulting in a saving of something like 40,000 kilocalories per year of each highland child, a saving which makes them capable of standing the hypoxia. It has also been found that sexual maturation is retarded by more than a year in the highlands. Levels and increase of LH don't differ from sea-level children; but in girls the high level of increase is found a year later than in girls of the sea level. It has also been found that the early modifications and adaptations of highlanders makes them have physical fitness comparable to athletes from lowlands.

The comparative studies of the permanent high altitude populations of Rio Pallanga at elevations of 4570 m and of the lowland populations at Lima (150 m above mean sea level), carried out by Krüger *et al.* (9), demonstrate that the placental weight is higher and the weight of the new born babies less at high altitude than near sea level. McClung (14) also records a correlation between the placenta and babies and altitude in two Peruvian Andean populations. There is thus a definite increase in the weight of the placenta in high altitude peoples as a kind of compensating mechanism for effectively increasing the volume and surface so as to bring about improved oxygenation of the foetus. Neonatal mortality of high altitude residents of the Peruvian Andes was studied by Mazess (12). The blood potassium and sodium in the high altitude children of Oroya (Peru) were studied by Carhuayo (2). Moncola *et al.* (15) studied the urinary steroids in men born at an elevation of 3000 m in Peru and residing at an elevation of 4267 m. The excretion of oestrogen seems to decrease markedly during pregnancy at an altitude of 4200 m; the estriol excretion is most affected (22, 23). Baker *et al.* (1) have given an interesting account of marked changes in the density of bones in high altitude residents of the Peruvian Andes.

4. Respiratory and Cardiovascular Systems

Frisancho (4), who recently made comparative studies of the pulmonary function and human growth in a residential population of the Quechua from Nunoa in Puna, at elevations 4000–5500 m on the Peruvian Andes, found a significant retardation of growth in stature. He also observed that despite their smaller stature, the high altitude men are characterized by relatively greater circumference of the chest. The socalled, 'barrel chest' of Tibetans and Andeans is well known. The thoracic volume is on an average 12.15 cc at elevations between 4000 and 4500 m, compared to 10.5 cc in the men at sea level. The corresponding differences in vital capacities are 5.35 and 4.92 litres respectively. In the north High Simién of the Abyssinian Highlands there are permanent residential populations at elevations ranging from 1500 to 3700 m. The chest dimensions of these people are related to the functional differences in their respiratory physiology, which may be traced to their long residence under the high altitude conditions. They are reported to have larger forced-expiration volume and vital capacity, only very slight polycythemia, higher packed-cell volume and raised systolic blood pressure than the lowland populations (6). The observations of Russian workers like Bryantseva on children born and raised at elevations of 1500–2000 m in Kirgizia also show a higher thoracic circumference and respiratory volume per minute. The enlarged chest size, the increased lung volume and the predominance of the right ventricle over the left, observed in high altitude populations, are attributed to an accelerated development during childhood and adolescence. Compared to the mean stature of 165 cm and vital capacity of 4.92 in the sea-level residents, men at elevations of 4000–4500 m have a mean stature of 162 cm and a vital capacity of 5.35. The activity of the heart is described as being highly economical in the permanent residents of elevated regions of 2000 m, suggesting a lower requirement of oxygen for approximately the same energy supply. It must, however, be remarked that all these features are by no means exclusive characters of the high altitude peoples, since athletes at much lower elevations are more or less similarly characterized.

Interesting observations have been made on the respiratory and circulatory systems of high altitude natives. The general superior performance of the high altitude natives, compared to that of new comers to mountains, is traced to the improved cardiac output, due to cardiovascular readaptation to the prevailing conditions of oxygen deficiency in the air. The investigations on the intake of oxygen and the body temperature of the Andean high altitude natives, resident at altitudes of 4000 m, carried out by Mazess (13), show elevated rest metabolism in them.

The pulmonary ventilation of men born and permanently living at high altitude is less intense in response to exposure to hypoxic conditions than in natives of sea-level regions. The investigations of Hultgren *et al.* (7) on the Peruvian Andean natives, resident at elevations of 3750–4328 m, show that the mean pulmonary arterial pressure is 22 mm of mercury (with a range of 11 to 17). The pulmonary arterial wedge pressure was, however, found to be normal. The calculated pulmonary arterial resistance shows 180% rise over that of the sea-level value. The difference between the arteriovenous oxygen content and the consumption of oxygen is slightly greater in high altitude residents than in sea level people, but the cardiac output remains normal. An interesting series of observations on the alveolar arterial oxygen gradient among the Andean natives have been made by Kreuzer *et al.* (8).

At an altitude of 4500 m, where the barometric pressure was 445.8 torr, native residents of the Bolivian Andes have an alveolar oxygen pressure of 46 torr, arterial pressure of oxygen was 45.1, with an arterial saturation of 80.1%. Ergometric exercises made in the highlands demonstrate that the maximum absorption of oxygen decreases with altitude, getting to be 1.4 litres per minute at an altitude of 7450 m in a healthy subject. Ventilation for a given amount of work, if it is light exercise, is independent of altitude, but increases at a given altitude until it arrives almost to the maximum of oxygen absorption. Any reduction in oxygen partial pressure in the inspired air results in a considerable disadvantage to maintain the diffusion capacity of the lungs. This is explained by the fact that the oxygen moves across the blood/gas barrier in a percentage, which depends on the difference between the partial pressure of the pulmonary alveolar gas and that of the pulmonary capillary blood. Therefore, any reduction in alveolar oxygen pressure from inhalation of air at low pressure should reduce the percentage of oxygen which moves across the alveolar membrane. If the level of exercise is high enough, the result may be arterial hypoxemia and an inevitable reduction and limitation in the capacity for work. Consequently, inhabitants of the highlands find themselves in condition of hypoxia and the persons temporarily exposed to altitude find that when the hypoxia is prolonged, there is an increase in the concentration of the hemoglobin which restores the oxygen transport capacity which the blood normally possesses. At greater altitudes it has been found that this restoration of transport capacity would not be completely utilized, because the maximum cardiac output is still lower than in controls at sea level. Probably the situation would be different after a prolonged stay in the highlands. The limitation noted in work capacity can be explained by the altered pulmonary diffusion, lessened cardiac output and the difficulty in achieving maximum pulmonary ventilation. It is admitted that there is a change in the chemical regulation of respiration, which is due not wholly to acid-base changes in the blood. This would not be sufficient even with the changes which respiratory alkalosis produces. It should perhaps be admitted that there exists an adaptation to prolonged hypocapnia, and which originally was maintained by the persistent hypoxic stimulation over the chemoreceptors.

Investigations with X-ray and autopsies have demonstrated in the Bolivian Andes that native highlanders have a moderate increase in size of the right cardiac ventricle. There is also an increase in relative weight of the same ventricle. Many native highland children have bradycardia, (an abnormally slow heart rate). Their electrocardiograms often demonstrate irregularities in systole and diastole.

After several days of stay in highlands it is found that cardiac output decreases, due principally to a lowering of systolic volume. Administration of plasma expanderers, in spite of the existence of hypovolemia, does not re-establish the systolic volume to sea-level values. It is possible that the cardiac function would be reduced by secondary hypoxia by the low oxygen pressure in arterial coronary blood or by a lessening of the flow of blood in these same arteries.

The reduction in cardiac output is on an average 16 litres per minute, when compared to the value of 23 litres per minute in the residents at sea level. The maximum cardiac frequency in the Bolivian Andean highlanders is always less than in the sea-level residents. The systolic volume is also lower than the value at sea level. The cardiac frequency is lowered by 13% but the flow of pulmonary blood rises by 60%.

The Bolivian native highlanders have adapted to the lowered oxygen pressure by increasing the haemoglobin and the number of red blood cells. With increase in altitude there is an increase in the quantity of circulating blood. This is the result of emptying natural storage places caused by hypoxia.

The miners working on the Peruvian Andes are grouped by Cosio (3) on the basis of the correlation between the haemoglobin value and the altitude at which they habitually reside and work. In the first group, the haemoglobin value is compatible with the hard life at high altitude. In the second group, the haemoglobin value is considered to be within the limits of adaptation to high altitude life and in the third group he describes what is known as 'Monge's disease', a condition characterized by severe increase of the haemoglobin value. From the point of view of the cardiovascular characters also the miners fall under three distinctive categories, viz. (i) those showing severe erythremia leading to pulmonary vascular congestion, (ii) those with severe pulmonary vasoconstriction, leading to pulmonary hypertension and dilatation of the pulmonary arteries and finally (iii) those with a combination of these two conditions. The telemetric investigations of heart of these miners, working at elevations of 4160–4800 m in Bolivia, carried out by Tejerina-Rayagada (24), have demonstrated changes of vagotonic character. Hyperurecemia is reported by Sobrevilla & Salazar (23) in people resident in a mining settlement, at an elevation of 4267 m in the central Andean Plateau. There is a marked rise of the serum rate, creatine and haematocrit, compared to the sea-level residents.

The blood viscosity in the haematocrit range of 40–78% does not offer resistance to the blood flow in the high altitude residents. In Bolivian highland native the haematocrit reaches very high values, as high as 80% at 4000 m, at which elevation the mean of lowland volunteers is only 61%.

The pH of the blood of highlanders is lower than normal for lowlanders; that is to say, the blood is less alkaline. The increase in H^+ ion is minimal, just sufficient to make the pH fall below 7.3. Similarly, arterial pyruvates and lactates decrease at altitude, but all three rise with intensification of physical exercise. Athletes in highlands show a lowering in organic phosphates, both in plasma and in red blood cells. At 5750 m, arterial blood from a reclining subject has a pH of 7.398.

The carotid bodies of native highlanders are enlarged due to the hyperplasia of the chief cells. Chemodectomas appear to be quite common in highland populations. There is microvacuolation around the neurosecretory vesicles of the chief cells in the enlarged carotid bodies. These cells perhaps secrete a polypeptide hormone called glomin. Increased size of the chemoreceptors seems to be associated with diminution of ventilatory response to oxygen deficiency.

The high altitude natives are generally characterized by longer circulation times, slightly greater cardiac output, higher total blood volume, as well as more blood in the lungs than in sea-level peoples. Significant differences in the relative thickness of the media of pulmonary trunk and the ascending aorta are described by Saldana & Arias-Stella (21) between the high altitude and lowland peoples. The permanent residents at elevations of 3414–4541 m have thicker medial coat of the pulmonary trunk and the ascending aorta than the sea-level people. This difference is attributed to the persistent mild pulmonary arterial hypertension in the high altitude people since birth. After the age of thirty, the media of the ascending aorta is thinner in the high altitude residents than in sea-level men, perhaps in relation to a lower systemic systolic pressure at high altitude. The same authors have also discussed the possible evolution of the elastic configuration of the pulmonary trunk in the high altitude natives.

Valuable data on the correlation between the basal metabolism, respiration and altitude are also available in the literature. Gill & Pugh (5) report, for example, that prolonged stay at elevations of 5800 m by the Sherpas of Nepal Himalaya have produced unexpected results. The pulmonary ventilation and the uptake of oxygen are similar both in the Sherpas and new comers (trained mountaineers) to the mountains. The Sherpas have, however, a lower alveolar ventilation, higher pulmonary arterial carbon dioxide pressure and lower pulmonary arterial oxygen pressure than the mountaineers. The rate of basal metabolism is 10% above the mean value expected from the sea-level standard. Among the Sherpas the mean metabolic rate is plus 21%. In the Sherpas, the hypoxic stimulation at extreme high altitude elicits only a relatively mild response, compared to that of the people from the plains, thus indicating long-term changes in cellular level (10).

5. Other Peculiarities

Hypofunctioning of the thyroid has been observed in the native inhabitants of nearly all highlands areas of the world. It is indeed known that exposure to hypoxia depresses the thyroid function. Thyroid hypofunctioning lowers the general metabolism and cellular oxidative processes. This condition approaches towards anaerobic glucolytic catabolism to meet the energy needs. It has been observed, especially in the Bolivian Andean highland natives, that the total thyroxine is lower than in lowland populations. It is also somewhat lower in highland native women than men. Experiments with radio-active iodine show that its uptake is somewhat higher in the Andean natives than in the sea-level population. It is reported that the protein-bound iodine decreases from 4.2 to 2.5 microgram per 100 ml of plasma, suggesting a better utilization of the circulating thyroid hormones or a blockage of its synthesis in the thyroid gland. While the intake of radioactive iodine is 74% at elevations of 1878 m, it is only 50.40% at 4060 m. In one area at 4540 m the uptake in 24 hrs was found to be 38.20%. The generally high cholesterol of the native of Bolivian highlands may perhaps be an expression of the functional irregularity of these glands. Certain interesting observations on the elimination of steroids through urine at elevations of 4200 m have been made in the Bolivian Andes.

It was found that the 17-hydroxycorticoids (17 OHC) and the 17-ketogenic steroids (17 KG), as well as the secretion of cortisol, increased significantly in persons exposed to hypoxia for two weeks. However, analyses after a longer exposure yielded normal results equal to the values at sea level. On the other hand, the 17-ketosteroids (17 KS) showed no variation, as normally happens when the adrenals are stimulated with oxogenous adrenocorticotropic hormones (ACTH). This would seem to suggest that the effect of hypoxia on the function of the suprarenals is different from that produced by ACTH. It was also found that the concentration of catecholamines in the plasma was normal in the first hours of exposure to hypoxia. Other studies on highlanders have shown that the response of the adrenals to ACTH is 30% less than in inhabitants of sea level, such that small doses of this hormone would be ineffective for highlanders. The highlanders need greater doses of endogenous ACTH for normal response. Exposure to high altitude causes an increase in plasma cortisol, probably due to an increase in secretion of endogenous ACTH. Various studies have shown that hypoxia would increase the secretion of ACTH, increasing the quantity of urinary steroids. They also show that the 17-ketosteroids have not changed in concentration. This can be explained as inhibition (by hypoxia) of the breaking of side chains of the steroids or by a lessening of the testicular fraction of the 17-ketosteroids. The latter explanation seems to be plausible. The transitory hyperactivity of the adrenals on exposure to high altitude could be affected. Since the adrenals play an important part in the tolerance mechanisms to lack of oxygen, this would be very important. Accepting that ACTH secretion increases with altitude, then hypoxia is a stimulus for the adrenal function so as to the increase concentration of ACTH in plasma.

The mechanisms will not be the same as what is called 'chronic hypoxia'. In this case, there is no response similar to the stimulus of exogenous ACTH. It is suggested that chronic hypoxia affects the possibility of using cyclic adenosinphosphate as a second hormonal messenger for ACTH and other hormones.

Comparisons have shown that highlanders have significantly higher concentrations (0.16 microgram per ml) of serotonin catecholamines than lowlanders. Nevertheless, there were no significant differences in epinephrine and nonrepinephrine. A case of insufficiency of suprarenal cortex which demonstrates the possibility of tolerance of chronic hypoxia without a significant polycythemia and with hypofunction of the suprarenal.

It has been found that glycemia in fasting is lower than that found in sea-level residents, and that there are no great differences in tolerance to intravenous glucose. In native highlanders glycemia is lower in basal conditions, but the return to normal values after administration of glucose and glycogen was faster and more marked. Half an hour after administration of glucose, immunoreactive insulin was significantly lower in native highlanders than in the sea-level control individuals. It is concluded that the lower levels of glycemia of the highlands is probably produced by a lower endogenous sensitivity to insulin and not precisely by a greater insulinemia. Tolbutamide tests yield results similar to that of natives of lowland. Insulin values are also similar to natives of the lowlands, although there is the possibility of a greater sensitivity to insulin at high altitudes, probably induced by hypoxia. In highlanders with the so-called chronic mountain sickness, there may be a retardation in secretion of insulin. Sensitivity to insulin is greater than at sea level, since several times it has been found that a dose of 20 to 30 IU has been able to induce therapeutic shock. However, incidence of diabetes itself seems to be no different at 3600 m than at other altitudes.

The volume of semen and the motility of spermatozoa are relatively low in native highlanders, but the concentration of spermatozoa is higher. Concentration of viable sperm does not differ from sea-level natives. Despite the differences in spermatogenesis, fertility is about the same as sea-level residents (53).

On rapid exposure to high altitudes, rapid lowering of testosterone production has been found, in the plasma. This does not impede the normal response of Leydig cells to adequate stimuli, although some studies have shown that the testicular response of highlanders to the gonadotropins is lower than in sea-level residents. The excretion of pregnanetriol in highlands is a third of that of the sea-level resident, further the increase that is found on stimulation with gonadotrophins is only 20% compared with sea-level response.

In human beings menopause appears later in highlands than at sea level. At elevations above 4000 m it appears between 16 and 18 years in native women, compared to 13 to 15 years in immigrants to highlands at 2750 m, (menstruation is produced at 13 years 11 months (56)). Acute exposure to high altitude causes changes in the hyoxia menstrual cycles; the LH does not increase in the middle of the cycle and the temperature turns monophasic, which confirms anovulary cycles.

Studies at 3200 m have shown that there is disturbance of kidney function on acute exposure, probably as a result of the effects of hypoxia on the central nervous system. After acclimatization the disturbance disappears. Studies on the volume and concentration of urine have shown a direct relation between diuresis and altitude, but above 3000 m the direct relation disappears. It was found that even though the volume of urine varies in spontaneous diuresis, it maintains some relation with the concentration. Of course, this is conditioned by characteristics of daily life in the high mountains, which decisively influence the balance of the water and the necessity of maintaining homeostasis.

6. Conclusions

We may summarize the information available so far by stating that the healthy native high altitude inhabitant is generally always in a state of mild pulmonary hypertension. This hypertension has an organic basis in the muscularization of the terminal parts of the pulmonary arterial tree. The rise in the pulmonary arterial pressure is greater in native children than in the adults. The elevated pulmonary arterial pressure of the highland native is really a persistence of the physiological pulmonary hypertension of the foetus. This induces the hypertroply of the right ventricle. In response to the mild pulmonary hypertension the pulmonary trunk shows also medial hypertroply. The normal modification of the elastic tissue of the media of the pulmonary trunk from the foetal to the adult type is delayed and there is an abnormal pattern of elastica. The pulmonary trunk of the highland native has, therefore, the general characters of the aorta of the sea-level resident. Some observers have recorded a diminished coronary blood flow in the native highlanders, but there is also increased vascularization of their myocardium.

In the highlander there is a redistribution of the blood flow, away from areas of the body with low oxygen extraction (like the skin) so as to increase the oxygen supply to the other parts. Total blood volume is increased due to the increased erythrocytes mass, with a decrease in the plasma volume. The fall in the cutaneous blood flow is associated with hyperkeratosis and hyper-pigmentation, leakage of melanin into the dermis and solar degeneration in dermis. The renal blood flow is also lower than in the sea-level residents. There is considerable variation in the changes observed in highland natives from different parts. The stages in the changes observed in different highland communities are: (i) initial exposure to high altitude or *accommodation;* (ii) prolonged residence or *acquired acclimatization;* (iii) native highlanders with *natural acclimatization;* (iv) partial adaptation as in barrel-chested highland natives; (v) full adaptation as in autochthonous animals; (vi) vascularlar loss of acclimatization or brisket disease; and (vii) respiratory loss of acclimatization or Monges disease.

Monge (16) believed that at least in the case of the Peruvian Andes, man seems to have become acclimatized for life at high altitude of 4575 m, especially to the extremes of low oxygen content of the air, high ultraviolet and cosmic ray activity, by having lived and reproduced for several generations. He assumed certain amount of genetic abilities to have given rise to the inherited structural peculiarities. According to him, different individuals can become more or less temporarily acclimatized to high altitude life, but true permanent acclimatization requires several generations of life and reproduction continuously at high altitude. When this acclimatization is successful, he concluded that the reproductive ability remains unimpaired. In a personal communication to one of us (MSM), Monge's son has recently questioned this view of the adaptation of natives at elevations above 4000 m. He has urged the need to revise his father's conclusions; it seems there is no case of such a permanent adaptation.

That the conclusions of Monge and others have only a limited validity is also demonstrated by a number of other observations. It is known, for example, that even in men who could be considered by Monge as fully acclimatized to high altitude, there are frequent cases of more or less acute pulmonary oedema, with physical symptoms of dyspnoea, cough, nausea, chest discomfort, haemophysis, cyanosis, tachycardia and hypertension on exposure to extreme elevations of 4267 m. As already said at the outset, pulmonary hypertension is common even in healthy natives of high altitude regions (18). It is generally believed that pulmonary hypertension is not capable of achieving an important part in acclimatization to high altitude, but only perhaps in association with hyperventilation, the extensive capillary bed of the lungs plays an important rôle in improving the arterial oxygenation. Nagasaka et al. (17) have also reported hypertension in the pulmonary circulation of men very well acclimatized to high altitude conditions, when they are exposed to simulated altitudes of 6000–7000 m. Mild pulmonary hypertension and increased pulmonary vascular resistance, greater than among adults, is reported by Simé et al. (20) in children between one and fourteen years of age, living at elevations of 4328 m (Cerro de Pasco) and 4541 m (Marococha). It would seem that the level of acclimatization of the native inhabitants of high altitudes may merely be somewhat better than that of new-comers from sea-level areas, but this is *by no means absolute or ever complete.*

The physiological and micro-anatomical changes in thoracic organs and in most other organ systems in the body, including the endocrine and reproductive systems, observed in the native inhabitants of highlands are generally described as adaptations. We do not however consider them at all as *normal,* but must regard them as pathological-physiological responses to oxygen deprivation. This view is substantiated by the observation that when these responses become even slightly exaggerated, as it does happen in numerous highland natives, they induce diseased condition, often with fatal results.

It seems that man has not lived long enough at high altitude to become adequately modified in a characteristic way that would differentiate him from sea-level residents. The process may no doubt be said to be in its incipient phases among the permanent residents of high altitude regions. Acclimatization depends evidently on the duration of residence at high altitude (*vide* Chapter XIV). How long has the human race colonized high mountain regions? Fortunately we have dependable evidence in answer to this question. Recent carbon isotope estimates of the human skeletons found at elevations of 3900 m in Lauricocha (Peru) and Chilca in South America indicate an approximate age of at least 7500 years. Although man has thus evidently lived since prehistoric times at high altitude, human colonization of the high altitude regions is by no means antecedent to the rest of the high altitude life. Quite unlike in the case of the greatest majority of hypsobiont organisms, which we have discussed in earlier chapters, the present day high altitude man represents relatively recent colonizations from lower elevations. These colonizations are also neither numerous nor extensive. The physiological basis of changes in hypsobiont animals like, for example, the llama, is entirely different. In them it does not lie in reducing the pressure gradient of oxygen from the ambient air to that at mitochondrial site as in man, but in a much more efficient extraction of oxygen in the tissues. The llama does not hyperventilate and has also a low haemoglobin concentration, but maintains an adequate arterial oxygen supply (with a low P_aO_2) on account of very high affinity of haemoglobin for O_2. These facts readily explain the relative incompleteness of the high altitude specializations in man. Considering all the peculiarities of the permanent residents of various high altitude regions in the world, the high altitude man must be described merely as more or less *acclimatized* to the extreme conditions, prevailing at high altitude, and not certainly *specialized* for high altitude environment. Ecologically the high altitude natives would really approximate to the hypsophile rather than to the hypsobiont type of organism. It is this lack of high altitude specialization that does not prevent peoples of the high mountain countries living at lower elevations in the plains. Though the highlanders may suffer various forms of temporary discomfort and may even have 'poor health', largely because of the excessive heat of the plains, they do not by any means suffer permanent injury, which can be strictly attributed directly to the low altitude and the excess of oxygen and higher atmospheric pressure of the plains. Likewise though man from the plains may suffer from acute mountain sickness on rapid ascent to high altitude, he recovers more or less completely on prolonged stay at high altitude or also on return to lower elevations. The cold hardiness of the hill peoples is also not an exclusive feature and not a part of their high altitude specializations.

7. References

1. Baker, P. T. & M. A. Little, 1955. Bone density changes with age, altitude, sex and race factors in Peruvians. *Hum. Biol.*, 37 (2): 122–136.
1a. Baker, P. T. & M. A. Little, 1978. Man in the Andes: A multidisciplinary study of high altitude Quechua. Stroudsburg (PA): Dowden, Hutchinson & Ross. pp. 303.
2. Carhuayo, D. D. 1965. Kalimia y natremia en ninos aparentements sanosque vivem en la altitud. *Biol. Soc. Quim. Peru*, 30 (2): 46–62.
3. Cosio, G. 1969. Mining work in high altitude. *Arch. environ. Health*, 19 (4): 540–547.
4. Frisancho, A. R. 1969. Human growth and pulmonary function of a high altitude Peruvian Quechua population. *Hum. Biol.*, 41 (3): 365–379.
5. Gill, M. B. & L. G. C. E. Pugh, 1964. Basal metabolism and respiration in men living at 5800 m (19,000 ft). *J. appl. Physiol.*, 19 (5): 949–954.
6. Harrison, G. A., C. F. Kurchemann, M. A. S. Moore, A. J. Boyce, T. Amju, A. E. Mourant, M. J. Gordber, B. G. Glasgow, A. C. Kopec, D. Tillis & E. J. Clegg, 1969. The

effects of altitudinal variation in Ethiopian populations. *Philos. Trans. R. Soc. London,* (B) 256 (805): 147–182.

7. Hultgren, H. N., J. Kelly & H. Miller, 1965. Pulmonary circulation in acclimatized man at high altitude. *J. appl. Physiol.,* 20 (2): 233–238, 239–243.

8. Kreuzer, E. S., S. M. Tenny, J. C. Mithoffer & J. Remmers, 1964. Alveolar arterial oxygen gradient in Andean natives at high altitude. *J. appl. Physiol.,* 19 (1): 13–16.

9. Krüger, H. & J. Arias-Stella, 1970. The placenta and the newborn infant at high altitudes. *Amer. J. Obst. Gynaecol.,* 106 (4): 586–591.

10. Lahiri, S. & J. S. Milledge, 1965. Sherpa (Himalayan high altitude residents) physiology. *Nature,* 207 (4997): 610–612.

11. Mani, M. S. 1974. Fundamentals of High Altitude Biology. New Delhi: Oxford & IBH Publishers, pp. 196.

12. Mazess, R. B. 1955. Neonatal mortality and altitude in Peru. *Amer. J. phys. Anthropol.,* 23 (3): 209–213.

13. Mazess, R. B., E. Picon-Reategui, R. B. Thomas & M. A. Little, 1969. Oxygen intake and body temperature of basal and sleeping Andean natives of high altitudes. *Aerosp. Med.,* 40 (1): 6–8.

14. McClung, J. 1969. Effects of high altitude on human birth. Observations on mothers, placentas and the newborn in two Peruvian populations. Cambridge: Harvard Univ. Press. pp. xvii + 150.

15. Moncola, F., E. Pretell & J. Corria, 1961. Studies on urinary steroids of man born and living at high altitudes. *Proc. Soc. exptl. Biol. Med.,* 108 (2): 336–337.

16. Monge, C. M. 1955. El concepto de acclimatacion. *An. Fac. Cien. Med. Lima,* 38 (1): 1–8.

17. Nagasaka, T., S. Ando, T. Takai, M. Hara, T. Satake & K. Tagaki, 1967. An analysis of EKG recorded by radiotelemetry in Mt Aconcagua and in a low pressure chamber at sea level. *Nagoya J. med. Sci.,* 29 (4): 377–384.

18. Penaloza, D., F. Simé, N. Banchero, R. Gambosa, J. Cruz & E. Marticorena, 1963. Pulmonary hypertension in healthy men born and living at high altitudes. *Amer. J. Cardiol.,* 1 (2): 150–157.

19. Pugh, L. G. C. E. 1963. Tolerance to extreme cold at altitude in a Nepalese pilgrim. *J. appl. Physiol.,* 18: 1234.

20. Simé, F., N. Banchero, D. Penaloza, P. Gambosa, J. Cruz & E. Marticorena, 1963. Pulmonary hypertension in children born and living at high altitudes. *Amer. J. Cardiol.,* 11 (2): 143–149.

21. Saldana, M. & J. Arias-Stella, 1963. The evolution of the elastic configuration of the pulmonary trunk in people native to high altitude. *Circulation,* 27 (6): 1094–1100; The thickness of the media of the pulmonary trunk and ascending aorta in high altitude natives. *ibid.,* 27 (6): 1101–1104.

22. Sobrevilla, L. A., I. Romero, F. Kruger & J. Whittembury, 1968. Low oxygen excretion during pregnancy at high altitude. *Amer. J. Obst. Gynaecol.,* 102 (2) (6): 828–833.

23. Sobrevilla, L. A. & F. Salazar, 1968. High altitude hyperurecemia. *Proc. Soc. exptl. Biol. Med.,* 129 (3): 890–895.

24. Tejerina-Rayagada, M. 1967. Telemetria cardiaca en mineros que trbajen entre 4150 y 4800 metros de altura sobre el nivel der mer. *Rev. Clin. Espan.,* 105 (4): 273–276.

25. Torrence, J. D., C. Lenfant, J. Cruz & E. Marticorena, 1971. Oxygen transport mechanism in residents at high altitudes. *Resp. Physiol.,* 11 (1): 1–15.

26. Zimina, R. P. 1964. Zakonomeri nosti vertikalnovo rasprostraniya Melkopitayushik. Izdatelstvo Nauka Mosow. pp. 1–156.

I Man in highland ecosystem: effects of exposure to high altitude

M. S. Mani

1. Introduction

This chapter presents a brief review of our knowledge of the effects of exposure of the resident of the plains to high altitude conditions on mountains. Although considerable attention has been paid to the study of the physiology of permanent residents of high altitude regions, the literature on the effects of high altitude on the sea-level residents and their reactions is even more extensive. The sea-level man at high altitude presents a number of unsolved problems of great fundamental importance in biology and medicine, often having close relation with military operations. Extremely valuable contributions to our knowledge of the reactions of man exposed to high altitude environment have been made in the course of numerous mountaineering expeditions to the Himalaya and Karakoram (118). Some recent work has also been undertaken by the Indian Army medical workers on the problems of troops operating at high altitude on the Himalaya. Quite contrary to the prevailing belief of those days, early experience of mountaineering on the Himalaya showed that the performance of the mountaineers, is fairly satisfactory up to an elevation of about 6900 m. Above this altitude was observed rapid and severe deterioration, marked progressive weakness, lethargy, failure to recover from fatigue, wasting of muscles and tissues, loss of appetite and a gradual decline of capacity for work. It is known, for example, that the strength of muscles in man at an altitude 2400 m is about one-third of that at sea level and at an altitude of 3400 m it is about one-fourth of the sea-level value. At altitudes between 2000 and 3400 m there is usually slight effect, but between 4000 and 5800 m the muscle fatigue is considerable and about 6100 m it becomes very pronounced.

As one ascends to elevations above 5800 m there follows a progressive worsening of physical and mental conditions. Although heavy loads can be carried at elevations higher than 5800 m, even slight muscular activity is followed by pronounced exhaustion. Anorexia and loss of body weight become acute. A variety of disturbances like dyspnoea, headache, nausea, vomiting, loss of appetite, aversion to food or abnormal appetite for certain foods, etc. begin to be observed. Mental impairment may be observed at elevations between 6700 m and 7900 m. Hypothermia, frostbite, snow blindness, splinter haemorrhage and other phenomena are frequent. Hypothermia results when the temperature of the human body falls below 35°C and proves fatal when the temperature of vital organs reaches 25°C by cardiac failure. Cooling also unfavourably affects the release of oxygen from haemoglobin. Hypothermia is associated with mild mental confusion or even delirious condition. It was also soon discovered that in the absence of previous experience on mountains and sufficient acclimatization to high altitude, man rapidly loses consciousness and dies at an elevation of 7500 m. Experimental exposures of human beings to simulated altitudes of 7500 m in low pressure chambers were found to cause unconsciousness in ten minutes and at a simulated altitude of 8230 m within three minutes. The Himalayan expeditions also found that breathing oxygen perceptibly slowed down the heart rate. Similar results have also been obtained by breathing oxygen at elevations of 3750–4328 m on the Andes. Some recent investigations have demonstrated that people, treated with acetazolamide, before ascent to an elevation of 5400 m on Mt. Logan, suffered relatively mild disturbances (89).

The pathological changes suffered by sea-level residents at high altitude are generally known as 'mountain sickness'. Acute mountain sickness is a syndrome, induced by exposure to hypoxia or low oxygen content of air at high altitude and leading after a brief time lag, to complex symptoms, including headache, insomnia, nausea, gastro-intestinal upsets, respiratory and circulatory upsets, neurohormonal disturbances, etc. Acute mountain sickness was first described in modern scientific literature by Father de Costa in 1590. In 1848 Albitskij made the important observation that when the oxygen of the air is reduced by 90%, the metabolic process of man is permanently raised, so as to compensate for the oxygen deficiency. When we consider the effects of exposure to high altitude on the inhabitants of the plains and their reactions, we observe that hypoxia is *par excellence* the dominant phenomena. The effects of atmospheric cold and high aridity and the action of ionizing rays are only of secondary importance and are indeed profoundly modified by hypoxia.

2. Hypoxia

The term hypoxia refers to deficiency of oxygen and includes also the reactions of the organisms to inadequate supply of the gas. An extensive literature has accumulated in recent years on hypoxia, particularly because of its importance from the point of view of military operations on the Himalaya since the Indian and Chinese border conflict and also because of its much greater significance in supersonic aviation. While some very valuable observations have been made in the field on highlands, the bulk of the recent investigations on hypoxia, with special reference to aviation medicine, has been carried out in the laboratory, inside low-pressure chambers, to simulate various high altitudes.

A perusal of the extensive literature shows that almost all experimental work has so far dealt with hypoxia in grand isolation from all other environmental factors, which influence organisms at high altitude. The results thus obtained do not, strictly speaking, reflect the reactions of animals to the effects of high altitude on mountains. As we have already stressed above, the action of low oxygen content of the air on animals is very profoundly modified by the interaction of atmospheric cold, and cold in turn influences the effects of hypoxia, so that the same organism reacts wholly differently to hypoxia at room temperature and hypoxia at low temperature.

i. *Hypoxia in Lower Animals*

Before considering the action of high altitude hypoxia on man, we may briefly refer to some of the interesting investigations on lower animals. It has been found that hypoxia is much less marked in the poikilotherm vertebrates than in the homoiotherms. The effect of the oxygen deficiency of the air at high altitude begins to be felt by human beings at pressures less than 350 mm of mercury, by monkeys at 300 mm, pigeons at 350 mm, cats at 270 mm, dogs at 250 mm, rabbits at 200 mm, but only at 100 mm by frogs. A number of observers have shown that *Rana esculenta*, *Bufo vulgaris*, *Lacerta viridis*, *Emys orbicularis*, etc. exhibit no appreciable symptoms of hypoxia, when exposed to simulated elevations of 3000 m. The investigations of Russian workers on the comparative physiology of the haematogenic functions, at an altitude of 2000–3700 m, show no appreciable reticulocytosis in the amphibian blood, though some slight increase is observed in the reptilian blood. While hypoxia is the primary stimulus for erythropoiesis in the homoiotherms, there is no such change in the fish and amphibia. Altland *et al.* (21) refer to Gordan as having observed in 1935 that only very immature erythrocytes are released in the peripheral blood in the amphibian *Necturus maculosus*, when exposed to high altitude hypoxia. High intensity hypoxia (breathing mixtures of 2.5–8.9% of oxygen), by mammalian standards a severe deficiency of oxygen, has no observable effect on the chelonians *Chrysemys marginata* and *Emys orbicularis*. It has been established that in these animals, breathing air with higher nitrogen content increases the respiratory activity more than the low-oxygen mixtures. According to Lumsden, quoted by Altland & Parker (21), the respiratory quotient of the turtle, due to hypoxia, is only slightly greater than in the rat, because the output of carbon dioxide is not reduced, while consumption of oxygen drops rapidly. The ability of the turtle to survive prolonged hypoxia, despite also the low oxygen consumption, seems to be due to the utilization of carbohydrates as an anaerobic source of energy. The blood of the turtle has also a much higher capacity for carbon dioxide than the mammalian blood and the erythrocytes of the turtle are characterized by much higher longevity. Experimental hypoxia in simulated altitudes of 13,716 m (with oxygen at 0.23 mm of mercury and gas pressure of 111 mm of mercury) fails to produce any change in the erythrocyte count, haemoglobin and haematocrit values or to induce reticulocytosis in these animals. The hypoxia induced by breathing 3–5% oxygen at 20–23°C alters the breathing cycle by reducing the rate and increasing the tidal volume, but not the volume of the expired air; oxygen consumption remains unaffected. In the poikilotherms, the atmospheric cold of the high altitude lowers the general

179

metabolism and as a chain reaction brings about a fall in the rate of the heart and of the pulmonary ventilation.

ii. *Hypoxia in Man*

The case is, however, different in man. The low atmospheric temperature at high altitude raises the heart rate in order to maintain the body heat. The observations of Bullard (52) have shown that, for example, shivering and oxygen consumption, heart rate and pulmonary ventilation in man are higher at 5°C than at 30°C, after exposure for thirty minutes and breathing only 10% oxygen. This is quite different from the poikilotherms, in which atmospheric temperatures of 36–38°C raise the respiratory activity, but do not bring about any change in the consumption of oxygen.

Even in the plains, any strong physical work leads in man to a certain degree of insufficiency of oxygen in the regional blood and causes tissue hypoxia, known as *hypoxia motoria*, depending on atmospheric temperature, barometric pressure and the duration of the physical activity. Maximum oxygen deficiency in man arises from violent exercise for one to three minutes, but recuperation is brought about by increased respiration and circulation. This hypoxic condition is greatly intensified even on slight physical exertion at high altitude.

In man, the effects and the severity of hypoxia at high altitude depend to a large extent on age and previous experience of exposure to hypoxia. The limited observations made on the Himalaya are fully corroborated by the data in the available literature. The observations of Lauer (114) and others have shown that the perturbations due to exposure to high altitude hypoxia are only very slight in children, less marked in the healthy adult, but pronounced in older people, in whom there are also anatomical impediments to adequate ventilation and cardiovascular regulation. The effects of hypoxia on man and his reactions are greatly influenced by earlier residence at high altitude and by previous experience of experimental exposure to oxygen deficiency. This is very well illustrated by the investigations of Russian observers on athletes on the Pamirs. Some of the athletes had five to seven years experience and were between 25 and 35 years of age, others were also experienced mountaineers of about eighteen to twelve years of experience of climbing and the rest were only primary sportsmen with five to seven years of athletic training. The observations on muscular strength, made both while climbing of the Lenin Peak (7134 m) and during the period of stay on the mountain for a period of forty days, demonstrate the existence of marked difference between the people who had previous mountaineering experience and others new to the mountain. The high altitude symptoms like exhaustion, insomnia, muscular weakness especially in the thorax and diaphragm, difficulty in breathing, etc. among athletes under training at high altitude on the Tien Shan are more or less marked, depending on previous experience of mountain climbing. The general symptoms are as a rule much less marked in the people who have trained previously on the mountain than in new entrants. On Mt. Terskol middle aged men showed increase in the respiratory volume per minute at an altitude of 2000 m, but the change was much less in younger men and in men trained earlier. Considerable amount of work along these lines has been done in Peru (130, 131).

The effects of hypoxia in man are very greatly influenced by the mode of its onset. Slow ascent of a high mountain by the plains people, allowing sufficient time for the body functions to get readjusted to the changing conditions, produces markedly less severe symptoms of hypoxia than rapid transport by car or by aircraft even to an elevation of 2400 m. When a resident of the plains is rapidly transported to high mountain regions, he usually suffers from some hyperventilation, tachycardia, malaise, respiratory dysfunction with slow and irregular and shallow breathing, pulmonary congestion with unproductive cough, more or less intense throbbing headache, some dizziness, perhaps also nausea, marked lassitude, insomnia and muscle cramps. Rise in pulmonary blood volume, often by 80%, results in acute pulmonary oedema at an elevation of 3660 m within about twelve hours. In the case of a gradual ascent of the mountain, most of these symptoms of altitude sickness not only appear at relatively higher elevations, but the symptoms are also often less acute than in rapid ascent.

The pathogenicity of high altitude hypoxia becomes evident in individual tissues as well as in the body as a whole. Disruption of gaseous exchange is always characteristic; the inadequate supply of oxygen to the tissues changes the basal metabolism and increases the oxygen requirement by the tissues. Hypoxia also causes an inadequate supply of the coronary circulation, producing a pronounced deficit in the need and supply equilibrium.

The effects of high altitude hypoxia on men may conveniently be considered under the following heads: (i) changes in the respiratory system (ii) changes in the cardiovascular system (iii) tissue reactions, including physiological and chemical conditions of the blood and other tissues and (iv) metabolism. Though dealt with separately here, the respiratory, cardiovascular, blood, tissue and other changes are on the other hand not independent, but phasic chain reactions to a fall in the oxygen supply to the tissues.

3. Respiratory Changes

The general difficulty in breathing felt by the man from the plains at high altitude must be considered as merely the first visible pulmonary symptom of deep-seated respiratory imbalance in the tissue level, the disturbance in the acid-base balance and other complex metabolic disturbances. The immediate affects of exposure to high altitude hypoxia in the sea-level resident are pulmonary hyperventilation, hypertension and pulmonary oedema. While there may be no significant change in the vital capacity, there is a more or less marked alteration in the volume per minute. At an elevation of 4000 m on the Himalaya, the rate of respiration falls on an average from 15 to 13 per minute in a healthy adult male, but at the same time the volume per minute rises from 7.4 to 8.91 litres. The ratio of pulmonary ventilation to the uptake of oxygen is greater at high altitude than near sea level. The oxygen cost of breathing is high. The mean alveolar pressures of carbon dioxide are found to be 20.7, 15.8 and 14.3 mm of mercury respectively at elevations of 6400, 7400 and 7830 m on the Himalaya (with barometric pressures of 344, 300 and 288 mm of mercury) and the corresponding values for oxygen are 38.1, 33.7 and 32.8 respectively (82). Increased differences in alveolar oxygen have been observed likewise even under simulated altitude exposures. The relation of pulmonary hyperventilation to high altitude, its causes and its effects have been extensively studied in recent years (155, 156).

It is now generally recognized that pulmonary hyperventilation is primarily a symptom, rather than a cause and is essentially in the nature of a reaction of the organism to compensate for the lack of oxygen in the blood and tissues. Hyperventilation at high altitude is controlled by the stimulus of the chemoreceptor organs along the aorta and the carotid sinus, which are considered to be sensitive to changes in the oxygen pressure in the arterial blood. It is also now established that the respiratory centre in the human brain, which controls the respiratory movements of the lungs, does not, however, react at sea level to the oxygen tension in the blood, but actually to the effect of changes in the tension of carbon dioxide. The pulmonary hyperventilation at high altitude leads, therefore, to a fall in the carbon dioxide pressure in the lungs, with also a corresponding fall in the blood (Fig. 13-1) so that brain really lacks the normal stimulus and is merely responding to the stimuli from the peripheral receptors, which are sensitive only to change of the oxygen tension. This leads to serious complications in man. Experimental investigations on pulmonary hyperventilation under high

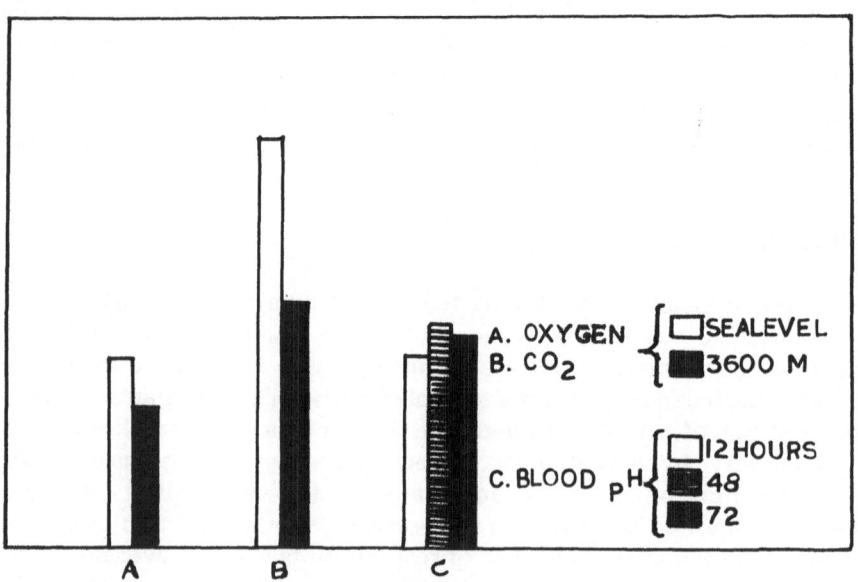

Fig. 13-1. Histograms of changes in the arterial blood characters on ascent of a sea-level resident to an elevation of 3600 m.

altitude hypoxic condition seem to have given semi-contradictory results; the problem is still not clearly understood. The lack of carbon dioxide stimulus as factor in hyperventilation is demonstrated by the observation (196) that breathing a mixture of air with 10% carbon dioxide increases the tolerance of cats to experimental hypoxia, under simulated altitude of 18,000 m. Others have however, stressed the importance of oxygen pressure as the important factor. Certain investigations have indicated that if the carbon dioxide level is maintained at the normal air value, the pulmonary ventilatory response to hypoxia is identical both at high altitude and sea level. It has also been shown recently (105) that the stimulus for ventilation by the inhalation of carbon dioxide is attributed to the effective pH in some regions, which are readily penetrated by carbon dioxide and less readily by other acids and thus tending to alter the acid-base character in such a manner as to restore its own pH to the normal. The acid-base changes are thus considered to be inadequate to explain the changes in the regulation of breathing during the action of high altitude hypoxia. The changes in the acid-base balance are to be traced to the accumulation of lactic acid. As the chemoreceptors in the carotid sinus react only to the fall in the oxygen tension, the result is pulmonary hyperventilation (28). The increased resistance to hypoxia, induced by inhalation of carbon dioxide, seems to be due to marked cerebral vasodilation and a corresponding increase in cerebral blood flow. In 1963 Riedstra (157) found that the hydrogen-ion acts as a specific chemoreceptor stimulus in hypoxia at high altitude. For arterial pressures of 100 mm of mercury, it stimulates the chemorecepter at pH below 7.35. In the range of excess acid the increase in fixed acid has two opposite effects on a critical value of arterial oxygen pressure, determined by the central pH and peripheral pH on the hypoxic stimulation of the peripheral chemoreceptors. Kellogg (105) believes, however, that the changes in the chemical regulation of breathing during the action of high altitude hypoxia cannot simply be explained in terms of the blood acid-base changes, which are not sufficient to compensate for the respiratory alkalosis.

Pulmonary hypertension has been reported at an altitude of 3500 m in men unacclimatized to high altitude, but in acclimatized persons only at an elevation of 4500 m. Pulmonary hypertension may be the primary symptom or it may also be followed by acute pulmonary oedema. Pulmonary hypertension arises from the thrombosis of the small pulmonary arteries. Pulmonary vasoconstriction, increased blood flow and polycythemia play a secondary rôle and any vigorous exercise is known to precipitate its onset even in natives of high altitude. The intensity of pulmonary hypertension in some adults on the Himalaya was higher at 3900 m than at 3000 m and this difference may be perhaps attributed to the duration and severity of hypoxia.

There seems to be some close relation between pulmonary hyperventilation, hypertension and oedema at high altitude. The general pathology and causes of the high altitude oedema have been recently investigated, both in man and a variety of experimental animals, on mountains as well as in low pressure chambers (144). A sudden rise of the pulmonary arterial pressure produces pulmonary oedema, without rise of the wedge pressure. Arias-Stella & Krüger (31) have made valuable contribution to our knowledge of high altitude pulmonary oedema and its pathology. They have also discussed the alveolar and bronchiolar zonal oedema, accompanied by hyaline membrane and thrombosis of the septal capillaries and small medium pulmonary arteries. Pulmonary oedema is attended by arterial oxygen desaturation, with hyperventilation and low carbon dioxide pressure. The disturbance in the acid-base balance leads to reversal of respiratory alkalosis into severe acidosis in pulmonary oedema.

There is no general agreement regarding the lowest altitude at which pulmonary oedema first appears in man. The altitude limits actually vary within very wide limits. While some workers have reported severe oedema in man at such a low altitude as 2400 m, others have observed it only at an elevation of 4000 m. Cases of acute pulmonary oedema in healthy adults, engaged in vigorous physical exercise at an elevation of 2400 m are on record (75). Roy *et al.* (161) have also reported pulmonary oedema and respiratory distress and infra-alveolar fibrine in man at elevation of 2440 m. Most cases of oedema apparently arise at altitudes of 3000–3500 m (24, 122). Indian troops, not formerly acclimatized to mountain warfare, developed acute mountain sickness and pulmonary oedema within 12–96 hours of reaching elevations of 3000–4917 m on the Himalaya at the time of the Indo-Chinese border conflict (161). The vulnerable altitude for the appearance of acute pulmonary oedema is generally believed to be 3350 m, at which alkalosis is induced by hyperventilation (118). Even at an elevation of 3000 m, the onset of acute pulmonary oedema may be precipitated by unaccustomed vigorous physical exercise (210).

4. Cardiovascular Changes

The respiratory disturbances are closely associated with cardiovascular upsets. Even at a moderate elevation of 1500–2000 m, there is often a marked acceleration of the blood flow, fall in venous pressure and a moderate rise in the permeability of the peripheral capillaries in some individuals. At a little higher elevation of 3000–3600 m, there is significant rise in the volume/minute of heart, rise in the velocity of blood flow, greater fall in venous pressure, rise in pulse rate, fall in peripheral blood volume of the cerebral flow, often by 40%, within the first twelve to thirty-six hours. With increased pulmonary ventilation, there follow increased polycythemia, pulse and respiratory rhythms, fall in arterial pressure and rise of the capillary permeability. The increased pulse frequency, the fall in the systolic pressure, diminution of the systolic volume of the heart from 119 to 103, retardation of the blood flow for the capillary vasodilatation and dilatation of the vessels of the pulmonary circulation are some of the striking changes commonly observed on the Himalaya. The Russian workers from the Pamirs have reported 4–6% reduction of vital capacity, pulmonary ventilation by 12% and fall of tachycardia by 8%. Acute high altitude hypoxia induces severe polycythemica, cardiac hypertrophy, altered cardiac output and adrenal cortical hypertrophy. The following table summarizes the comparative data on the mean values of pulmonary-cardiovascular indices during brief stay of the sea-level resident at altitudes of 3500–3600 m on the Himalaya:

Table 13-1. Changes in the physiological indices during brief stay of sea-level residents at moderate elevation on the Himalaya.

Indices	Plains (150–166 m)	3500–3600 m on the Himalaya	
		3–5 days	5–11 days
Respiration/ minute	20	25	27
Pulse rate	76	89	93
Systolic arterial pressure	118	109	103
Diastolic arterial pressure	77	72	66

An excellent review of literature on the relation between the cardiovascular changes and altitude may be found in Mirrakhimov (125). Jackson (103) has devoted considerable attention to the study of the human heart under high altitude conditions. He observed that with increase in altitude, there is a more or less pronounced faster and deeper respiration, alveolar ventilation and increased cardiac output and a quickening of the heart rate. More haemoglobin also passes through the lungs per minute. There is an increase in the haemoglobin level but a fall in plasma volume (Fig. 13-2). Pulmonary arterial constriction and increase in the pulmonary artery pressure, more pronounced electrocardiogram with dyspnoea, cyanosis cough and pulmonary oedema result from anoxia at high altitude. The changes in the human heart, as a result of exposure to high altitude conditions on the Andes in Bolivia, are discussed by several workers. Marked increase in the pumping activity of the heart has been observed as a result of the high altitude hypoxia (15, 47, 124). This is to be attributed to changes in the pulsating velocity of blood and to the increase of work of the left ventricle by about 25%.

An interesting investigation by Novikova & Kapelko (142a) on the myocardial functioning under the action of high altitude hypoxia at simulated altitude of 6000 m, has shown that there is an increase by 33% in the maximum contraction of the right ventricle and a 25% increase in the mass of the ventricle. At elevations of 7300 m, there is very marked hypertrophy of the right ventricle. The mass of the right ventricle increases by 30%. The maximum systolic pressure of the ventricle rises by about 58%.

Differences in the cutaneous circulation at sea level and at high altitude have also been studied (119). There is an increase in the tone of resistance and capacitance at high altitude, which becomes particularly marked when the skin temperature is higher. In other words, when the cutaneous circulation is raised as a circulatory adjustment under the action of high altitude

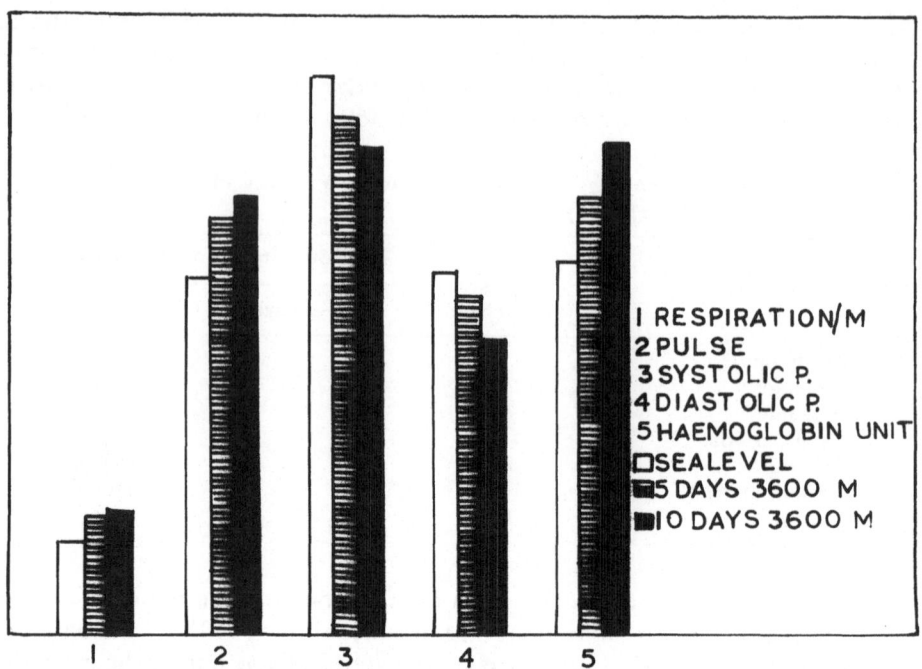

Fig. 13-2. Some of the important physiological indices of the sea-level resident, before and after ascent to moderately high altitude on the Himalaya.

hypoxia, this change occurs. Certain ultra-structural peculiarities of the heart muscle have been described in dogs, experimentally exposed to various stages of hypoxia (193). The cardiac hypertrophy has been traced to synthesis of nucleic acid proteins in rats, exposed under experimental conditions to prolonged hypoxic atmosphere (121). In the first week of exposure of experimental rats to hypoxia of simulated altitude of 7000 m, the myocardium changes in such a way that the right ventricle is more resistant than the left. Fibrotic changes in the hypetrophied right ventricle have been described after exposure to simulated altitude of 5000 m for four weeks. The physiological load of physical exercise in man, under conditions of high altitude, was measured in terms of heart rate and oxygen consumption, from sea level to 4000 m, both before, during and after exercise (101).

5. Changes in the Blood

As is well known, one of the principal functions of the blood in the vertebrates is its rôle in respiration. The disturbances of the respiratory function of the blood under the influence of high altitude hypoxia have been extensively investigated (104, 134, 186).

Observations at simulated high altitudes, under experimental conditions inside low pressure chambers, have established that there is marked increase in the ratio of the mean blood volume to body (104). Pugh has also studied the changes in the blood volume and the concentration of haemoglobin at high altitude. On the Himalaya and Karakoram, after a stay of about eighteen weeks at elevations between 4000 and 5800 m, the mean blood volume is found to be less than at sea level by 9%. On further stay of 3–6 weeks and 9–14 weeks, there is a fall in plasma volume by 27%. The erythrocyte volume and total haemoglobin rise by 49% above the sea-level value. It was also found that the haemoglobin concentration in the men studied by Pugh (*loc. cit.*) on the Himalaya does not equal that of the Andean residents at altitude of 5340 m. In experiments with dogs, exposed to simulated altitude of 5300 m for sixteen weeks, Vogel *et al.* (198) have reported a 59% reduction in the plasma volume. Blood volume changes in man are reported during three weeks at high altitude. Simultaneous measurements of the plasma volume and cell mass in polycythemia of high altitude in thirteen residents of Cerro de Peso (4330 m) in Peru have shown mean values of 33.4–9.44 ml/kg of body weight for plasma and 51.7 ± 16.33 ml/kg of body weight for erythrocyte volume. An interesting observation has been made by Barbashova (43) on the correlation between the osmotic resistance of the erythrocytes and altitude in white rats, experimentally exposed to altitude of 3900 m. Destruction of erythrocytes, observed under high altitude hypoxia in experimental dogs (85), is attributed to the increased haemolytic activity in part and to the lowered resistance of the new erythrocytes in part.

Even in the early days of mountaineering on the Himalaya, an increase in the haemoglobin index with altitude was observed. The increase of haemoglobin and the erythrocyte may amount to 11%. The ascent of high mountain is associated with more or less marked changes in the composition of the blood proteins, not merely of the haemoglobin, but also there is increase of ascorbic acid, glutathion, cytochrome-C, cytochrome oxidase, succinic oxidase, succinic hydrogenase, cyclophorase, histaminase, carbonic anhydrase, catalase, globin oxidase, adenosintriphosphates, etc. In rats experimentally exposed to simulated altitude of 1000–8000 m for several months, there is marked increase of the total haemoglobin from 0.75/100 of the body weight at sea level to 1.1/100 of body weight at an altitude of 4000 m and maximum of 2.2/100 of body weight at an altitude of 6000 m. The haematocrit level at sea level was 45%, at an altitude of 4000 m 60% and at 8000 m 85% (192).

The important changes which arise in the composition of the blood proteins of man before, during and after the ascent of high altitude are summarized in Table 13-2 from partly unpublished data collected by Russian workers on the Pamirs (Fig. 13-3).

The important phenomena of erythropoiesis, during prolonged action of high altitude hypoxia, is discussed by a number of workers (35, 38, 45, 77, 127, 192). The formation of erythrocytes is in inverse proportion to the value of the oxygen capacity of the blood *ab initio* of the exposure to the hypoxic atmosphere. The fall in the plasma erythropoietin level, with continued hypoxia, is not in any way associated with decrease in the rate of erythropoiesis. Guineapigs, experimentally exposed to simulated altitude

Table 13-2. Blood proteins at high altitude.

Blood protein fraction	850 m	At 4000 m		On return to 850 m
		1 month	40 months	
Albumin gr%	66.2	55.7	56.7	54.8
	4.8	4.7	4.3	4.2
Alpha$_1$ globulin	2.9	5.5	5.8	5.7
	0.2	0.4	0.4	0.4
Alpha$_2$	22.9	3.9	5.3	6.4
	0.2	0.4	0.4	6.5
Beta globulin	13.2	16.4	14.3	16.6
	1.0	1.2	1.1	1.3
Gamma globulin	14.8	18.5	18.1	16.5
	1.1	1.5	1.4	1.3
Total protein	7.3	8.5	7.6	7.7

of 3000, 3660 and 6000 m, show more young reticulocytes at 6000 m than at 3000 m. The guineapigs also show increase in erythrocyte level under experimental exposure to high altitude hypoxia. The effect of high altitude hypoxia on the platelet count is described by some authors (36). Haemato-crit, haemoglobin and erythrocytes levels increase rapidly for two weeks in experimental guineapigs and then more slowly with homeostasis between seventh and tenth weeks. There is also an increase of the peripheral reticulocytosis during the first two weeks of exposure to high altitude. An abnormally high level of the erythrocyte 2,3-diphosphoglycerate is found in man at altitudes of 3100–3400 m and this is considered to be an important contributory condition in the development of polycythemia at high altitude (70).

Comparative studies of the haematopoietic function in man and some other experimental animals like frog, chelonian, snake, hamster, rat and cat have shown that in the sea-level resident there is a marked rise in the haematies and haemoglobin on ascent of high altitude (Fig. 13-4). This

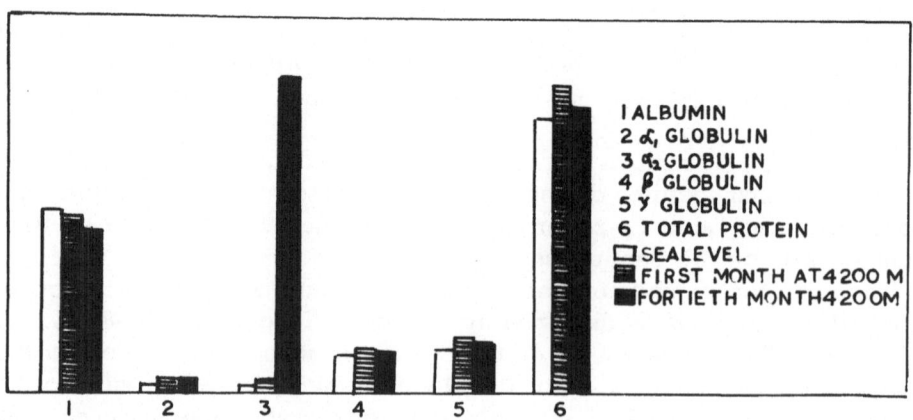

Fig. 13-3. The percentage changes in the blood proteins of sea-level resident on ascent to an elevation of 4200 m on the Pamirs.

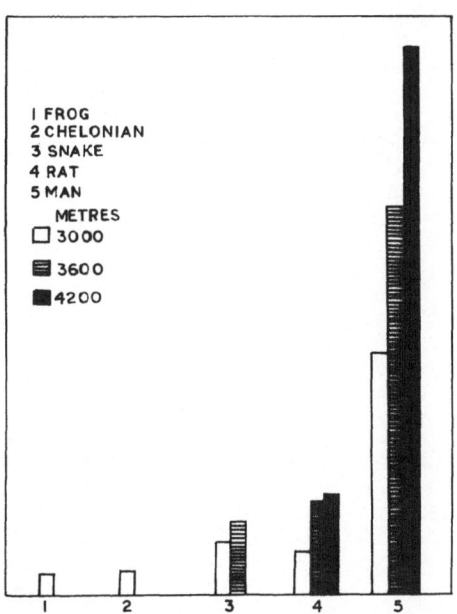

1 FROG
2 CHELONIAN
3 SNAKE
4 RAT
5 MAN

METRES
☐ 3000
▤ 3600
■ 4200

Fig. 13-4. Histograms showing the comparative intensities of erythropoiesis in some important vertebrates and in man at different elevations.

function is much less marked in the poikilotherms than in the homoiotherms. There is also higher erythropoietic activity in hibernating mammals than in the non-hibernating ones when exposed to high altitude hypoxia. There is no marked change in the spleen at 3000 m, but there is a significant increase at 6000 m.

The data on the rabbit blood bone-marrow haematogenesis and the peripheral blood in relation to the duration of residence at high altitude, collected by Andreyeva, Chernova and other Russian workers may indicate the changes likely to be found in the human being exposed to the high altitude hypoxia (Table 13-3).

The increase in blood viscosity and the higher haematocrit, under exposure to high altitude hypoxia, generally bring about thrombotic conditions. Thrombophlebitis in calf, pulmonary thromboembolism, pulmonary thrombosis and infarction, hemiplegia from cerebral thrombosis have been observed frequently in mountaineers, at elevations generally between 5800 m and 7930 m. Thrombosis is believed also to be favoured by the pronounced dehydration of the body at extreme elevations.

Blood coagulation is affected as a result of high altitude hypoxia. There is an increase in plasma fibrinogen level and fibrinolytic activity. Platelet adhesiveness has also been observed. The disturbances in blood coagulation lead to fibrin thrombi in pulmonary capillaries and arterioles (in case of pulmonary oedema).

Some reference may be made here to the regulatory activators of the respiratory function of the blood (83). It has been observed that under conditions of high altitude hypoxia, experimental dogs reveal a sharp

189

Table 13-3. The haematogenic index (bone medulla) under high altitude hypoxia, summarized from the data collected by Russian workers.

Haematogensis	Sea level	Number of days of residence at high altitude				
		3	5	10	15	20
Primordial cells	1.94	1.34	1.03	0.73	1.53	0.83
Myeloblasts	0.88	0.61	0.73	0.30	0.61	0.90
Promyelocytes	2.27	3.80	3.13	2.40	3.81	3.03
Myelocytes (Neut.)	5.97	6.30	7.81	9.53	10.21	8.90
Myelocytes (eosinophile)	0.75	0.31	0.36	0.13	0.33	0.83
Neutral Metamyelocytes	4.72	5.41	3.41	4.20	6.26	6.60
Eosinophile Metamyelocytes	0.28	0.27	0.36	0.26	0.26	0.30
Lymphocytes	13.8	9.60	8.30	11.13	9.21	12.70
Plasma cells	0.60	0.52	0.41	0.20	0.06	0.60
Proerythroblasts	1.18	1.03	0.91	0.37	0.62	0.93
Erythroblast basophile	3.11	3.46	2.93	1.60	2.06	2.40
Erythroblast acidophile	14.76	17.30	19.73	15.75	17.80	11.13
Leucocyte/ erythrocyte ratio	2.12	2.78	0.97	1.36	1.41	2.71

increase in the haemoglycosis in the first days of exposure, thus leading to dissociation of the haemoglobin-oxygen. Though the haemoglobin and glutathione amounts remain unchanged, there is a fall in the alkaline reserve. The respiratory function of the blood becomes subsequently activated, because of an increase in the haemoglobin level up to 16 gr% in the course of ten days. The fall in the alkaline reserve becomes normal from sixty to ninety days. In 100–110 days, with the fall in air temperature, the levels of haemoglobin and of the alkaline reserves fall and there is a rise in the haemoglycosis activity. Comparable data in man are not available.

6. Neurohormonal Changes

The partial deprivation of oxygen and the changes induced in the blood at high altitude set up a series of disturbances in the endocrine and nervous functions as chain reactions. The results of some of the investigations carried out on experimental animals are likely to be of importance in the case of man also, and may be briefly reviewed here.

Utilizing the technique of labelling with radio-active isotope of phosphorus, Tiisala (190) studied the response of endocrine organs in new-born and adult rats exposed to high altitude hypoxia. The hormonal factors in the tolerance of rats, exposed experimentally to simulated high altitude conditions, investigated by de Bias (63) are important. It was found, for example, that the survival of adenalectomized rat is less than that of the normal rats, at a simulated altitude of 8470 m. In case of mild hypoxia of moderate high altitude, all hormone doses are protective and bring about a normalization of the functional state of the cortical neurons. The administration of hydrocortisone in 0.6 mg doses per 100 gr of body weight has no adaptive effect. This is explained by Fedorova (13) by the effect of hormone on the functional stability and liability of nerve formations. The changes in the adrenal cortical function induced by high altitude condition have also been studied by some workers (129).

Extremely interesting results have been obtained by several recent workers regarding the functioning of the thyroid gland under the action of high altitude hypoxia, particularly in experimental rats (26, 93). It seems that while no visible pathological changes may be observed in the thyroid of animals and of man under exposure to high altitude conditions, there is nevertheless well marked hypofunctioning of the gland. The thyroid hypofunctioning leads to a series of complex changes like retardation of growth, fall in fertility and in severe cases even total sterility (66, 86, 181). Certain marked abnormalities often found in experimental rats, under hypoxic conditions of simulated altitudes of 7620–8230 m, can only be explained on the basis of thyroid and other endocrine imbalance (66). Nueurosecretory shifts in the blood, induced by high altitude hypoxia, must perhaps explain the inhibition of the cortex, leading to hypnotic state at altitude of 6000–7000 m.

A number of workers have investigated the effects of high altitude hypoxia on the functioning of the central nervous system. The problem is of considerable importance in aviation medicine. Some of the results obtained, particularly by Russian workers, have far reaching significance but unfortunately for security reasons, the data are not generally available to us.

In experimental cats, subjected to simulated altitude of 18,000 m, some of the functional and morphological changes induced by hypoxia involve the appearance of tonic convulsions within 5–10 seconds of exposure, followed by muscular asthenia (196). The maturation of the brain at an altitude of 3660 m has been investigated by Heim (96). The oxygen tension in the brain of the dog, exposed to high altitude of 12,000 m and breathing air, fell by half; the tension in the subcortex was two-thirds of the normal (109). The physiological indices of the brain and muscle state during high altitude exposure (3200 m) in rats were observed by some Russian authors (11). Other Russian workers have also recorded marked increase in nervous activity in man during prolonged exposures, up to 20 hours, to rarefied air, corresponding to altitude of 12,000 m, but breathing pure oxygen. Alifanov (13) has studied the relation between high altitude hypoxia and the motor and sensory chronoxy of man. The same author (15) has also described the changes in the motor conditioned reflexes in man under high altitude hypoxic condition, in simulated altitude of 5000 m, for 20–25 minutes and breathing air or to altitude of 10,000 m and breathing oxygen. The subjective symptomatology and cognitive performance of man at high altitude were studied by Stampfer et al. (78). During high altitude hypoxia, a progressive diminution of the sensitivity of the eye to light has been observed. Experiments have also been conducted by Agadzhanyan and his collaborators (5,

6, 7, 8, 9) on the motor balance and stability of organisms under high altitude conditions. Prolonged restriction of movements at high altitude was found to significantly lower the efficiency of all experimental animals. Agadzhanyan (3) has also dealt with the extinction of the conditioned motor electrodefensive reflexes under the influence of hypoxia of simulated altitude of 6000 m. He found chaotic excitation when the reflex is extinguished, instead of the gradual diminution. Excitation was also found to occur earlier at an altitude of 8000 m than at 6000 m and the signal activity of the cortex is completely upset. With the fall in the visual acuity and of the brightness sensitiveness of the eye at high altitude, it has been found that the severity of headache also rises in man (74). It seems that in the course of evolution in the vertebrates, there is an increasing importance of the higher sections of the brain in the reactivity of the animal to high altitude hypoxia (114). The disastrous effects of severe high altitude hypoxia on human brain include fatal brain oedema. Numata (143) has given a brief account of the high altitude brain oedema, which appeared at an elevation of 6400–7260 m on Mt Makalu in the Himalaya. The Japanese mountaineer Tagaki seemed dull when camping at 7260 m (snowing), scanty of words and had little appetite. When he was brought down to 6530 m, he took milk-tea but vomitted it promptly. Takagi of the Chiba University Japanese Mountaineering Expedition to Nepal 1971 died of the effects of oedema of brain. The symptoms included delirium, unsteadiness of legs, incontinence of urine, unconsciousness, convulsion and death. The oedema develops slowly and the symptoms appear equally gradually. This is one of the very serious manifestations of acute effects of high altitude sickness. It is not relieved by administering oxygen or by return to lower altitudes. Oedema is usually accompanied by some haemorrhage in the brain. The best treatment is considered to be administration of corticoids and diuretics like lasix.

7. Metabolism, Growth and Reproduction

As explained above, the pulmonary, cardiovascular, blood and neurohormonal functional upsets are not isolated effects of the influence of hypoxia of high altitude, but are always associated with disturbances in every organ, in the general metabolism, growth, fertility and reproduction.

Some of the more important of these disturbances may be noticed here. Disturbances of the reflex correlation between salivary and renal activities have been observed at altitude of 2000–3000 m on the Himalaya. Observations also on the Himalaya have indicated the existence of a close relation between uropoiesis and altitude. It has been observed that the relation between spontaneous diuresis and an altitude of 2000 m is greatly accentuated at a higher altitude of 3000 m. Inhibitory changes in the salivary glands have been observed by Russian workers at altitudes of 3000–4000 m, in intestinal glands at 8000 m, in bile formation at 6000 m, in digestive enzymes at 4000–5000 m and in the cytochrome anhydrase at 10,000 m. (simulated altitude). The incidence of gastric and duodenal ulcers at high altitude, reported by Nemeh & Villegas (139) would seem to indicate hypersecretion and hyperactivity induced by hypoxia. While some workers found that the functional capacity of the liver is not seriously affected by hypoxia at an altitude of 3450 m in man, others have described in experimental rats exposed to simulated altitude of 6000 m, a 40% increase in the liver fat within the first ten days and increase almost by twice the original value later. Male rats, exposed under experimental conditions to simulated altitude of 3540 m for ten days, show 19% decrease in the consumption of oxygen by liver in the case of forty-day old animals and 30% fall in eleven-day old animals. Burgos et al. (53) investigated the changes in the rat hepatocyte, after exposure to high altitude of 4700 m for two weeks. The total esterified fatty acids were more than three times that of the control rats, but the glycogen was much less. Hepatocytes with high cytoplasmic lipid droplets were found near the central vein of the hepatic lobules. There was also a fall of the endoplasmic recticulum. The hypoxia of high altitude severely interferes with the normal development of hepatocytes, which thus remain enzymatically immature. A 50% reduction in the liver glycogen content and 25% fall in the blood sugar value are reported in mice, experimentally exposed to simulated altitude of 3800 m for one month (48). This would show a lowered rate of conversion of glucose to carbon dioxide. The influence of high altitude hypoxia on the glucocorticoid and endotoxin and induction of hepatic enzyme were studied by Berry et al. (46).

Reference has been already made to the changes in the blood proteins under the influence of high altitude hypoxia. It is well known that almost 80% of the proteins are synthesized in the liver and the change in the blood proteins are to be traced to the fall in their synthesis in the liver.

High altitude hypoxia alters the consumption of vitamins. It is known, for example, that the system of dehydrogenases and flavoproteins constitutes an oxidative process of cellular coenzymes, thiamine, riboflavins and nicotinic acid. The system of cytochromes and oxidases, with catalisers like ascorbic acid, forms oxidative process. The hypoxia of high altitude produces severe disturbances in this system, so that the consumption of the vitamins is profoundly affected, often setting up a condition of acute hypovitaminosis. The results regarding the relation of ascorbic acid are, however, partly contradictory. While some workers have observed a fall in the ascorbic acid, others have reported a rise and still others believe that there is only a redistribution. There is in man acidoalkalopenia and much elimination of urine during ascent of high altitude and there is also increased elimination of vitamin B_1 through urine. Mirrakhimov (125, 126) has discussed the close relation between high altitude hypoxia and the vitamin metabolism.

Some workers report observations on protein catabolism in experimental rats at simulated altitude of 4300 m (106). Increased synthesis of proteins and RNA in the brain of experimental rats at simulated altitude of 7500 m is also reported (120). Maximum protein synthesis was observed in the cerebral cortex on the fortieth day and this increase is considered to be an adaptive reaction of the animal to high altitude hypoxia. An increased rate of biosynthesis and release of catecholamines in the brain of rats, exposed to hypoxia of simulated altitude of 10,000 m, has been observed by some workers (146). The excretion of proteins in the urine of mountaineers on the Himalaya was investigated by Rennie & Joseph (156a). In investigating the amino acid catabolism under high altitude conditions, Whitten *et al.* (206) found that rats ate less food after forty-eight hours exposure to simulated altitude of 5486 m than the control ones. There was also an increase in transminase activity and urea cycle enzyme activity. The former is due to caloric restriction; part of the increased capacity for amino acid catabolism under the influence of acute hypoxia is due to caloric deficiency.

The carbohydrate metabolism in man is very profoundly altered under high altitude conditions. It has been repeatedly observed that after oral administration of blood glucose, lactose, pyruvate, plasma inorganic phosphate and plasma potassium, the concentration of glucose is lower in the blood in both arteries and veins at an altitude of 4540 m than at sea level. The normal utilization of energy by the myocardial cells is upset by high altitude hypoxia, resulting in increased production of high energy phosphates, parallel with the rise in the cardiac output. The phosphorylization process is also impaired at an altitude of 7925 m. There is an increased utilization of proteins, increased energy requirement and at the same time a decrease in the protein synthesis. The injurious effects of high altitude hypoxia seems to be really due to reduction in the level of ATP in the tissue rather than damage to the ATP-producing mechanism.

The excretion of aldosterones seems to be related to the retention of potassium during respiratory alkalosis in mild hypoxia (37, 172). During the first week of exposure to simulated altitude of 4300 m there is an increase in the serum level of chloride, phosphate, proteinate and calcium and a fall in sodium, potassium and magnesium.

The hypofunctioning of the thyroid, to which reference was made above, is closely correlated with retardation of growth, fall in fertility, sterility, abortion, etc. Gordan *et al.* (86) found that the primary abnormalities in rats, exposed under experimental conditions to simulated altitudes of 7620–8230 m, lead to retarded growth. Adrenal cortical insufficiency must also be considered as offering partial explanation of retarded growth and lowered fertility (34, 181). It has been found that generally the growth retardation by high altitude hypoxia is more pronounced in the male rat than in the female (62). Russian workers have, however, reported that the ovarioles of the experimental rats are far more sensitive to high altitude hypoxia than the testes. The retardation of growth occurs in the female in the first ten months of discontinuous exposure to hypoxia, but the decrease in the body weight is not as great as in the male. Cattle, ram and domestic cat show signs of reproductive failure, as a result of the action of altitudes of 3350–4560 m, but sheep, acclimatized since perhaps the seventeenth century to high altitude, do not show any loss of fertility. The failure of the male fertility is attributed to exfoliation of the germ cells into the ducts of the epididymis before maturation. The spermatozoa are outnumbered by immature cells. Azoospermia, decreased sperm motility and deviations of pH have also been observed (131). Abnormal spermatozoa have been described in rabbits, which had been exposed daily for 6–22 hours for two weeks to high altitude hypoxia. It is believed that tissue oxidation may perhaps explain the retardation of fertility in rats, four generations of which were exposed to high altitude hypoxia. This view is confirmed by the observation that with rise of oxygen by 11% the rats started reproducing normally. The observation of Russian workers that the excretion of the 17-ketosteroids is 20% lower at high altitude than at sea level is of interest, when it is recollected that these are products of the androgen function and the testicular hormones and in the male they provide the esteroid synthesis of the suprarenals, which also account for almost two-thirds of the testicular esteroids. Pronounced irregularities in the oestrous cycle and the impairment of the fertility have been described in rat exposed to simulated altitude of 3800 m. Massive fatal lesions have developed in the placenta of albino guineapig, when exposed to a simulated altitude of 3900–4575 m (65).

8. Action of Ionizing Radiations

Although considerable attention has been paid in recent years to the action of ionizing radiations on organisms in general and on man in particular, we really know very little about the effects of ultraviolet radiation. The action of the ultraviolet rays on organisms is complex and involves various organs and tissues and functions. In so far as high altitude biology is concerned, our main interest lies in the action of the ultraviolet rays on the human skin.

Nearly all the experimental work on the effects of ultraviolet on human skin seems to have been confined to the skin on the arm and the back. It is, however, generally recognized that different parts of the human skin differ greatly in their sensitivity to ultraviolet rays at high altitude.

The exposure of the unadapted human skin of the white races to the noonday sun during summer produces within about half an hour visible sunburn, early pigmentation in some people and the emission of a peculiar odour from the skin and finally leads to the formation of vitamin D from sterols. In the course of three or four days, late pigmentation and even blisters may arise. The minimum erythemal dose or the solar radiation required to produce barely visible sunburn differs under different conditions. The human skin reflects about 20–40% of the visible part of the solar spectrum and absorbs the rest. It is known that it absorbs considerable ultraviolet.

The characteristic skin odour develops on exposure of the skin to the ultraviolet, only when the surrounding air is cool enough, as at high altitude, so as to produce no sweating. The odour is believed to be due to some breakdown products of nucleic acids or other proteins or of lipid conversion under the action of the ultraviolet rays.

Within a few days, the sunburn turns to permanent pigmentation and blisters arise. Prolonged exposure causes elastosis in some people. Early aging of the skin on the face and the lower arms and even skin cancer have been known to develop under the action of ultraviolet.

The data from laboratory spectroscopic studies, combined with observations on solar and ozone absorption, have shown that the maximum reaction occurs under natural conditions, when exposed to wavelengths in the neighbourhood of 307 μm of sun and skylight. We do not, however, know precisely much about the sensitivity of the human skin to wavelengths between 313 and 334 μm.

While relatively so little is known about the effects of ultraviolet on the human skin, we are almost completely ignorant about the influence of various other factors operating at high altitude on the action of ionizing radiations. There is also considerable confusion and contradiction in the conclusions arrived at by different workers. While some workers like Graveskii & Konstantin (87) consider that the high altitude hypoxia offers no protection against the action of ionizing radiations, others (55), assert the protective action of high altitude hypoxia. This view is strongly supported by certain known facts. The radiochemical effects on tissues are mainly formation of radicals HO_2, H_2O_2; the tissue PO_2 has an important influence on the formation of these radicals. Increase of PO_2 would ordinarily increase radiosensitivity, but fall of PO_2 occurs at high altitude. Therefore the effects of radiation are reduced. As HO_2 is a strong oxidizing agent and forms in presence of oxygen, hypoxia is likely to offer some measure of protection against radiation effects. It is also known that reducing agents retard biological action of radiation. According to Braun (50), high altitude causes an additional stress in the action of ionizing radiations. Animals chilled under high altitude hypoxic condition are said to exhibit increased tolerance of irradiation. Prolonged hypothermia is reported, however, to result in increased mortality on exposure to ionizing radiations at high altitude. High altitude hypoxia has been observed to suppress the essentially harmful reaction of irradiation, thus producing a protective antiradiation effect, somewhat similar to that produced by cystamine. The prolonged action of ultraviolet at high altitude must probably explain the high prevalance of cataract of the eye in the permanent residents of Tibet and the inner Himalayan valleys. Here is one of the most important aspects of high altitude that needs detailed investigation in the field.

9. Psychic Effects

It seems that in all earlier investigations on man at high altitude a great deal of over emphasis has been laid on the rôle of hypoxia and other external conditions. It would be wholly erroneous to assume, as indeed most workers have done, that man's reactions to high altitude, the diverse and complex symptoms of mountain sickness and the process of human acclimatization to high altitude are exclusively or even largely governed by the external factors like the low oxygen content of the air. Quite unlike in the case of various experimental animals, which have so far been studied both in the field and in the laboratory with special reference to hypoxia, human reactions do not end with physiological and metabolic cell-tissue level effects, however,

196

complex these prove to be, but extend to a much wider range of emotional phenomenon. There is increasing evidence to suggest that even among the lower animals like rats, widely used in experimental investigations stresses of the nervous system introduce newer and wholly unexpected complicating factors, which very profoundly influence the process of acclimatization to high altitude conditions. In man these stresses not only prove to be of far greater importance than in the lower animals, but may even completely outweigh almost all other factors.

The prolonged sense of discomfort of life under wholly unaccustomed and inhospitable conditions at high altitude, the continued physical exhaustion, restricted movements, the monotony and other associated factors result in very considerable emotional imbalance and stress, which set up complex chain reactions. (Fig. 13-5). The altered metabolism as a reaction to the high altitude hypoxia induces neurohormonal changes, which involve endocrine imbalance. This has, as a chain reaction, a considerable disturbing influence on the normal functioning of the higher nervous system. The mental inhibition due to life under extreme and prolonged discomfort upsets the normal endocrine equilibrium. The mental condition affects the endocrine functions and these latter in turn affect the mental condition. The functioning of the human mind, in other words his thoughts of the sense of discomfort, act as a *new* determining factor, to an extent so far wholly unsuspected by experimental biologists. It is well known, however, that the neurosecretory shifts in the blood caused by high altitude hypoxia give rise to hypnotic state at elevations of 6000–7000 m. Even at relatively moderate elevations the excitability of cerebral cortex is raised, thus acutely influencing the nervous processes and increasing the concentration in time of the process of inhibition-excitation. There is a marked intensification of the excitatory process at elevations of 2000–30000 m and a protective inhibition at 3000–5000 m. This is perhaps to be attributed to the close correlation between the functional state of the neurohormonal shifts and the endocrine substances in the blood, which transmit nerve excitations. At least in part, the emotional stress arises not merely from the conditions prevailing immediately in the high altitude environment, but it is due directly to the fact that man is essentially a gregarious animal. The social isolation, the forced separation from his associates, the loneliness, the compulsory throwing together of the same companions, the confinement in limited space, his acute anxiety to get back to normal life and the frustration at being prevented (though often necessarily by deliberate choice) and delayed, which are inevitable in any sort of prolonged stay at high altitude on mountains, adversely affect his mental state and add to his irritation.

The theatre of human evolution, like indeed that of every other terrestrial animal, has been in an *environment of green vegetation*, so that the human eye has what may be described as 'rest metabolism' and may be said to be in the 'ground state' when only acted upon by green colour. The action of all other colours, if prolonged, naturally upsets this ground state. The excessive glare and the intense blue at high altitude, even without having to consider the injurious effects of the ultraviolet part, hurt the human eye and set up

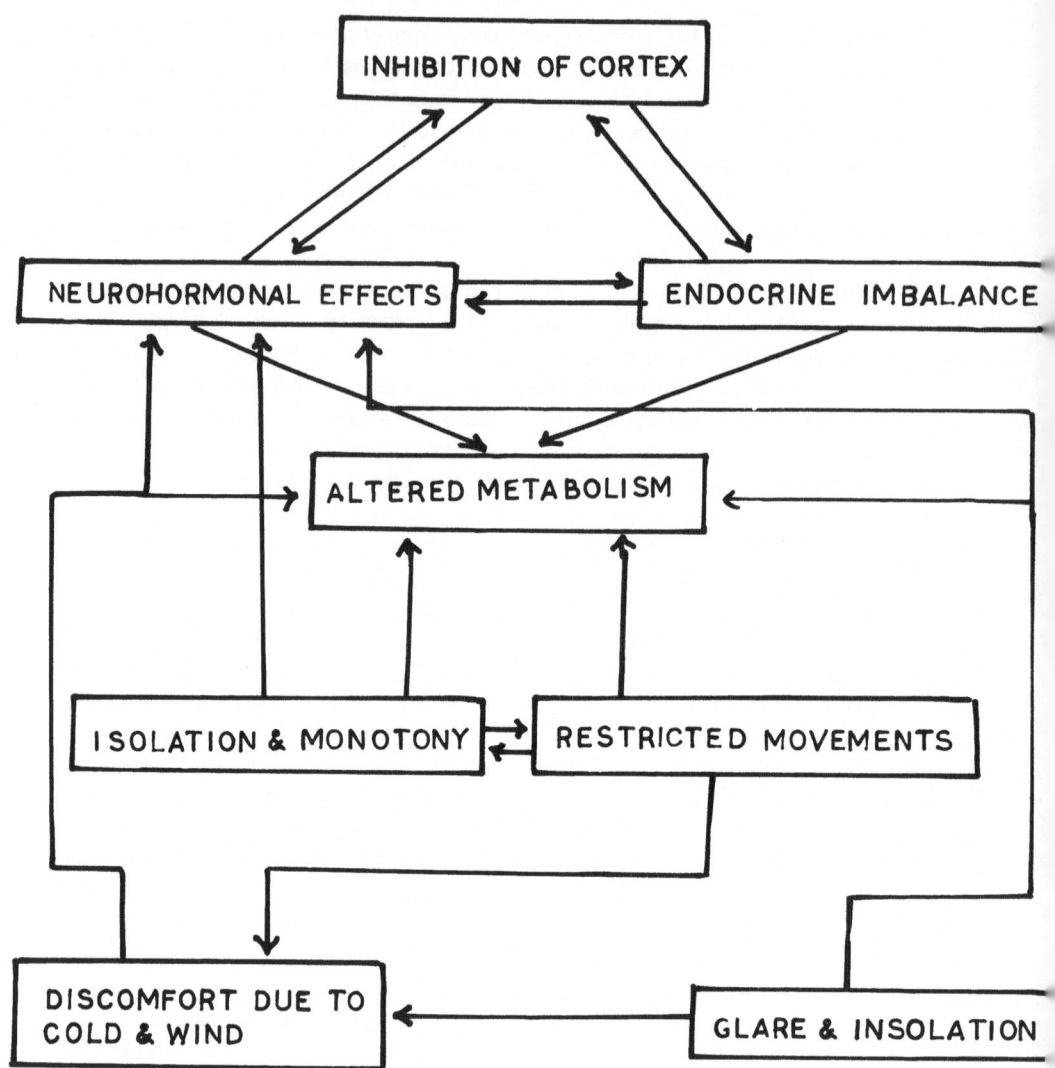

Fig. 13-5. Diagrammatic representation of some of the complex interacting factors, which produce psychic reactions as chain effects in a sea-level man exposed to the extremes of high altitude environment.

chain reactions in the higher nerve centres. This effect is quite distinct from and must not be confused with the well known symptoms of snow blindness of mountaineers and the all-too common disease, viz. cataract of the eye among the permanent residents of high altitude regions in the Himalaya and Tibet. The lack of green shade thus sets up unsuspected emotional stresses in man. More or less pronounced visual disturbances and nervous headaches, which are typical symptoms of the prolonged exposure to intense glare and excessive blue, completely disappear and recovery is rapid and total on his return to the shade of green grassy meadows and trees. It is the common experience of sea-level visitors to high mountains that it is difficult to estimate distances as readily as in the plains, partly because of the thinner air and consequently everything is exceptionally sharp and clear and this effect is greatly exaggerated by emotional upsets, so that things appear to be anything between a few metres to several kilometres away.

Human beings have the capacity of tolerating considerable atmospheric cold and indeed the major phase of human evolution as *gregarious species* must be attributed to the cold of the Pleistocene times. Modern man functions, however, best under moderate cold, so that prolonged excessive cold of the high altitude, notwithstanding the effective protective clothing, aggravates his already acute sense of discomfort. The constant and high velocity wind on mountains also further contribute to increase the chilling effect of atmospheric cold.

The cumulative effect of these complex factors of emotional stresses is to give rise to diverse, usually vague, but nevertheless more or less pronounced and sometimes curiously contradictory symptoms, which may perhaps be called 'high altitude neurosis', for want of a better expression, and which are at present usually but erroneously attributed wholly to the deficiency of oxygen in the air at high altitude. Some of the symptoms of the high altitude sickness are, strictly speaking, not high altitude but emotional sickness. The most usual sequence of these symptoms observed on the Himalaya begin with a progressively marked aversion to food, followed by a gradual loss of appetite. There is then a more or less marked depression of mental alertness and fall in coordination, incidence of irrelevant and discordant talk, tendency to shout and lose temper under slight or indeed no provocation or pass into long fits of silence. Indifference and pronounced loss of initiative supervene. There is often also considerable loss of ability to concentrate the mind on anything in particular. The reflexes may or may not be delayed. There may be marked drowsiness or also on the contrary unaccountable and sustained insomnia. There is always more or less marked visual disturbances and tendency to exaggerate the size of objects, confuse their contours and colours, misjudge their distance and see moving objects where no motion exists, especially in the middle distances. The story of the abominable snowman finds its natural explanation in this state of mind (116). Hallucinations may be more or less pronounced or may also be completely absent. There are often intervals of uncontrollable aggressiveness and an irresistable desire to 'run away from it all', followed equally unaccountably by brief intervals of lucid and logical behaviour, normal mental faculties and perfectly normal and gentle manners. There is a constant and vague fear of impending and inevitable doom. While no doubt all or nearly most of these symptoms are essentially and primarily emotional and have their parallel in the well known arctic sickness of Siberian winters (140, 141, 142) hypoxia naturally accentuates the emotional basis. There is some unconfirmed evidence suggesting that these symptoms of high altitude neurosis are not severe and may even be absent in individuals, who have sufficiently acclimatized at altitude of 3000–4000 m. Regular physical activity at high altitude would also appear to partially reduce the severity of the neurosis.

It must, therefore, be concluded that human reaction and behaviour at high altitude are certainly not simple and cannot also be satisfactorily interpreted purely on the basis of observations on experimental animals studied inside low pressure chambers. Like every other organism, the human being also reacts to various individual factors, to the complex of the interactions of separate factors, to the sum total of all these factors and to his own reactions to these factors complex, but quite unlike other animals, all these reactions are finally very profoundly modified by his own mental state. We, therefore, deal here with a complex of closely correlated and integrated effects, reactions, mental setup, etc., which are at present only extremely imperfectly understood. The scientific investigations of these problems are greatly complicated, because the behaviour of no two individuals is identical under the same given set of external experimental conditions. We are, therefore, completely unable to lay down rule-of-thumb observational methods in the study of man at high altitude. We do not by any means assert here that the various symptoms enumerated above are always evident; some or most of them may altogether be absent and the sequence of their appearance may also be different from what we have described here. Some of the symptoms persist more or less or may even disappear altogether after a short interval. The most remarkable and without doubt the constant feature of the psychic effects at high altitude is indeed their utter *unpredictability*. It must be evident that if they are readily predictable, they would perhaps cease to be psychic effects. It is these considerations that should make us hesitate in drawing hasty conclusions from experimental data on animals. While these results are no doubt extremely interesting and even instructive, they are certainly not decisive when applied to human beings. *Man reacts as much or perhaps even more to his inner environment as to the outer. The external environment no doubt influences his internal environment, but he largely creates the latter himself by his mental condition – by his thoughts* (we need not be afraid of bringing them into the field of science). The inner environment thus created by him may often present no measurable relation at all to the outer environment and nearly in all such cases the result is pathological upsets of various kinds.

10. References

1. Ackerman, E., J. Baldes & R. G. Slabaugh, 1962. Effect of exercise and simulated high altitude conditions on xanthine oxidase activity in rat livers. *46th Ann. Meetg. Atlantic City NJ Fed. Proc.*, 21 (2): 221.
2. Adrogue, H. J., J. C. Etcheverry & A. C. Taquini, 1970. Hemodynamic response to hypoxia in dogs with special reference to ventilation. *Acta Physiol. Latinoamer.*, 20 (1): 2–12.
3. Agadzhanyan, N. A. 1956. Ugashenie uslovnykh dvigatelykh elektrooobornitelnykh reflekslvo v usloviinakh razrehnnoi atmosfery. *Zhur. Vassh. Nervn. Deitl.*, 6: 260–268.
4. Agadzhanyan, N. A. 1961. Some questions of the study of flight work. A review of foreign literature. *Voenno-med. Zhur.*, 10: 72–76.
5. Agadzhanyan, N. A., L. L. Karpova & G. V. Machinskii, 1967. Dvigatelnvi rezhimi i vysotnoya ustoichivost organizm. *Voenno-med. Zhur.*, 11: 60–64.

6. Agadzhanyan, N. A. & G. Machinskii, 1968. Vlyanie dlitelnoi gipokinezii na visotnuyu ustoichivost belykh krys. *Kosm. Biol. Med.*, 2 (1): 25–28.

7. Agadzhanyan, N. A., M. M. Mirrakhimov, V. A. Isabaeva, V. K. Vasilev, G. A. Davydov, Zh. S. Dubinina, A. F. Zavodovskii & I. Kovaleva, 1968. Osobennosti izmenenniya gaszoobmena u gortsev i u zhitelei ravniny pri vypolnenii fizicheskii nagruski. *Sb. nauch. Tr. Med. Inst.*, 47: 17–22.

8. Agadzhanyan, N. A. & A. V. Sergienko, 1970. Perenosimost ostroi gipoksi posle razlichnykh srokov preleyvniya v srede povyshennym soderzhaniem uglekisloty. *Dokl. Akad. Nauk. SSSR* (Biol.) 19 (2): 487–489.

9. Agadzhanyan, N. A., M. I. Vakar, V. A. Smirnov, I. Chernikov & A. I. Shaposhnikov, 1961. Technique for measuring pulmonary ventilation during respiration against elevated pressure at high altitudes (Trans.) *Fiziol. Zhur. SSSR*, 47 (6): 77–79.

10. Agostoni, A. 1967. Acid-base disturbance in pulmonary edema (human). *Arch. Intern. Med.*, 120 (3): 307–310.

11. Aidaraliev, A. A., V. A. Berezovskii & M. D. Dzhunushev, 1970. Fiziologicheskie pokatateli sostoyaniya golonogo mozga i myshtsy v protsesse akklimatizatsii k uslovixam vysokogorya. *Fiziol. Zhur SSSR*, 56 (9): 1203–1214.

12. Alexander, A. F., D. H. Will, R. F. Grover & J. T. Reeves, 1960. Pulmonary hypertension and right ventricular hypertrophy in cattle at high altitude. *Amer. J. Vet. Res.*, 21 (81): 199–204.

13. Alifanov, V. N. 1956. The effect of hypoxia and low barometric pressure on the motor and sensory chronoxy in man. *Biull. eksptl. Biol. Med.*, (Transl.) 41 (1): 29–32.

14. Alifanov, V. N. 1960. The effect of hypoxic hypoxia and low barometric pressure on human electrocardiogram (vector analysis). *Biull. eksptl. Biol. Med.* (Trans.) 50 (10): 29–33.

15. Alifanov, V. N. 1961. Changes in motor conditioned reflexes in man in hypoxic hypoxia and low barometric pressure. *Zhur. Vysshei Nerv Deiatelnosti I. P. Pavlova* (Transl.) 11 (4): 28–33.

16. Altland, P. D. 1946. Effect of discontinuous exposure to 25,000 ft simulated altitude on body weight and reproductive system of immature rats. *Anat. Rec.*, 96: 27.

17. Altland, P. D. 1949. Effect of discontinuous exposure to 25,000 ft simulated altitude on growth and reproduction in albino rat. *J. exptl. Zool.*, 110 (1): 1–15.

18. Altland, P. D. & B. Highman, 1957. Effects of high altitude exposure on dogs and their susceptibility to endocarditis. *J. aviat. Med.*, 9: 92.

19. Altland, P. D., B. Highman & B. D. Nelson, 1968. Serum enzyme and tissue changes in rats exercised repeatedly at altitude: Effects of training. *Amer. J. Physiol.*, 214 (1): 28–32.

20. Altland, P. D., B. Highman & J. Roshe, 1959. Effects of altitude on dogs with valvular heart disease, tolerance and pathologic effects of acute and chronic exposures. *Arch. Pathol.*, 69 (1/6): 475–486.

21. Altland, P. D. & M. Parker, 1955. Effects of hypoxia on the box turtle. *Amer. J. Physiol.*, 180 (2): 421–427.

22. Altman, O. L. & D. S. Ditmar, 1961. Blood and other body fluids. *Fed. Amer. Soc. exptl. Biol.*, 1–191.

23. Altukhov, G. V. & N. A. Agadzhanyan, 1961. Higher nervous activity in man during prolonged exposure to rarefied air. *Zhur. Vysshel. Nerv. Deiatelnosti im I.P. Pavlova*, (Transl.) 11 (4): 7–11.

24. Alzamora-Castro, V., Guillermo Garridolecca & C. Battilana, 1961. Pulmonary oedema of high altitude. *Amer. J. Cardiol.*, 7 (6): 769–778.

25. Anthonison, N. R. & H. J. Smith, 1965. Respiratory acidosis as a consequence of pulmonary oedema. *Ann. intern. Med.*, 61 (5): 991–999.

26. Antumnez, C. & A. B. Houssary, 1961. Action de la hipoxia sobre la function tiroidea. *Rev. Soc. Argentina Biol.*, 37 (5/6): 112–119.

27. Archibald, E. R. 1964. The carbon dioxide response curve of the dog at sea level and at altitude. *US Air Force Tech. Doc. Rep.* AMRL-TR, 64-145: 1–152.

28. Ardashnikova, L. I. 1959. Znacheni razlichnykh khemoresepternykh oblastei v regulyatsii

dykhaniya pri nedostatki krisloroda. *Tr. Inst. Normal i patol. Fiziol Akad. Nauk. SSSR*, 3: 64–73.

29. Ardashnikova, L. I. 1968. O meckhanizmakh izmenenii arteralnogo davleniya i chastoy serechnych sokrashenii pri ostroi gipoksicheskoi gipoksii. *Byul. exptl. Biol. Med.*, 65 (3): 25–29.

30. Arias-Stella, J. 1969. Human carotid body at high altitudes. *69th Progr. Abstr. Amer. Assoc. Pathol. Bact.*, San Francisco, 150

31. Arias-Stella, J. & H. Krüger, 1963. Pathology of high altitude pulmonary edema. *Arch. Pathol.*, 76 (2): 147–157.

32. Arias-Stella, J. & M. Saldana, 1962. The muscular pulmonary arteries in people native to high altitude. *Med. Thoracalis*, 19 (5): 484–493.

33. Arias-Stella, J. & M. Saldana, 1963. The terminal portion of the pulmonary arterial tree in people native to high altitudes. *Circ.*, 28 (5): 915–925.

34. Armstrong, H. G. & J. W. Heim, 1938. The effect of repeated daily anoxemia. *J. aviat. Med.*, 9: 92–96.

35. Atherton, R. W. & P. S. Timiras, 1970. Erthropoietic and somatic development in chick embryo at high altitude (3800 m). *Amer. J. Physiol.*, 218 (1): 75–79.

36. Atkins, P. & R. E. Lempke, 1970. The effect of hypoxia on the platelet count. *Brit. J. Surg.*, 57 (8): 583–586.

37. Ayers, P. J., R. C. Hurter, E. S. Williams & J. Rundo, 1961. Aldosterone excretion and potassium retention in subjects living at high altitudes. *Nature*, 191 (4783): 78–80.

38. Baciu, I., C. Opaciu & M. Dorftieu, 1961. On the direct and reflex action of hypoxia in the nervous regulation of erythropoiesus. *Rum. med. Rev.*, 5 (1): 73.

39. Bacq, Z. M. & P. Alexander, 1955. Fundamentals of radiobiology. London: Butterworth's Scientific Publications.

40. Balazov, E., P. Masiar & V. Balaz, 1969. Obsah ninhydrinpositivnych latok v sredei i svale potkanov pri aklimatizacii na vysokohorske prostereidie. *Fysiat. Reumatol. Vesstn.*, 47 (1): 16–22.

44. Barbashova, Z. I. 1957. Response to hypoxia in animals with cerebellar damage. *Izv. Estev.-Nauchn. Inst. Imenie P. F. Lesgofta*, 1957 (28): 159–168.

42. Barbashova, Z. I. 1960. Specific and nonspecific lines of acclimatization to hypoxia. Acad. Sci. USSR. Problems of evolution of physiological functions. (Transl.).

43. Barbashova, Z. I. 1963. O korrelyasii mezdu rezistentnostyu organizma i osmoticheskoi rezistentnostyu eritrositov. *Fiziol. Zhur. SSSR*, 49 (5): 626–631.

44. Barbashova, Z. I. 1967. Studies on the mechanism of resistance to hypoxia: A review. *Intern. J. Biometeorol.*, 11 (3): 243–254.

45. Barbashova, Z. I., I. Griegoreva, V. V. Ermilova & Z. G. Fomia, 1959. Influence of the nervous system on hypoxic erythrocytosis. *Fiziol. Zhur. USSR* 45 (7): 103–111 (Transl.).

46. Berry, L. J., D. S. Smythe, L. S. Colwell & P. H. C. Chu, 1969. Influence of hypoxia, glucocoticol and endotoxin on hepatic enzyme induction. *Amer. J. Physiol.*, 215 (3): 587–592.

47. Bershtein, S. A. & Z. N. Aitkhozhina, 1970. O gemodinamicheskikh sdvigakh pri razvitii pressornogo reflekse s karotidnykh sniusov u usloviyakh ostroi gipoksii. *Izv. Akad. Nauk Kaz. SSSR* (Biol.) 3: 56–61.

48. Blume, F. D. & N. Pace, 1967. Effect of translocation to 3800 m altitude on glycolysis in mice, *J. appl. Physiol.*, 23 (1): 75–79.

49. Boning, D. 1970. Wirkungen eines akuten Sauerstoffmangels auf die Blutenelektrolyten-konzentrationen bei hohenangepassten und nichtangepassten Menschen. *Pfügl. Arch. Physiol.*, 314 (3): 217–230.

50. Braun, H. 1962. The effects of ionizing radiations under the influence of hypoxia and hypothermia. *10th intern.* Congr. Radiobiology Book Abst. Canada, 1075: 185–186.

51. Buettner, K. J. K. 1969. The effects of natural sunlight on human skin. In: Biologic Effects of Ultraviolet radiation with special emphasis on the skin. Ed. F. Urback. Oxford: Pergamon Press, p. 237.

52. Bullard, R. W. 1961. Effects of hypoxia on shivering in man. *Aerospace Med.*, 32 (12): 1143–1147.

53. Burgos, M. H., E. O. Zangheri & I. M. Parishii, 1970. The rat hepatocyte after exposure to high altitude: Structural and chemical changes. *Quart. J. exptl. Physiol. cog. Med. Sci.*, 55 (1): 54–58.

54. Butler, P. J. 1970. The effect of progressive hypoxia on the respiratory and cardiovascular system of the pigeon and duck. *J. Physiol.*, 211 (3): 527–538.

55. Bychkovskaia, I. B. & G. K. Ochinskaia, 1960. Protective effect of hypoxia at various radiation doses. *Biofizica*, 5 (4): 468–473 (Transl.).

56. Cain, S. M. & J. E. Dunn II, 1965. Increase of arterial oxygen tension at high altitude by carbonic anhydrase inhibition. *J. appl. Physiol.*, 20 (5): 882–884.

57. Chopra, K. L. & N. D. Menon, 1969. Reversal of symptomatic high altitude pulmonary hypertension and cardiac failure among temporary residents in Himalaya. *Indian Heart J.*, 21 (1): 37–47.

58. Coehour, J. K. 1960. Erythropoietic stimulatory factor production in pubescent mice after a single exposure to hypoxia. *Proc. Soc. exptl. Biol. Med.*, 105 (3): 673–675.

59. Consolazio, C. F., LeRoy O. Matoush, H. L. Johnson & T. A. Daws, 1968. Protein and water balances of young adults during prolonged exposure to high altitude (4300 metres). *Amer. J. clin. Nutr.*, 21 (2): 155–161.

60. Cotes, P. Mary, 1961. Bioassay of erythropoietein in mice made polycythemic by exposure to air at reduced pressure. *Nature*, 191 (4793): 1065–1067.

61. Cunningham, W. L., E. J. Becker & F. Kreuzer, 1965. Catecholamines in plasma and urine at high altitudes. *J. appl. Physiol.*, 20 (4): 607–610.

62. Dalton, A. J., B. F. Jones, V. B. Peters & E. R. Mitchell, 1945. Organ changes in rats exposed repeatedly to lowered oxygen tension with reduced barometric pressure. *J. nat. Cancer Inst.*, 6: 161–185.

63. De Bias, D. A. 1962. Harmonal factors in rat's tolerance to altitude. *Amer. J. Physiol.*, 205 (5): 818–820.

64. Delaquerriere-Richardson, L., S. Forbes & E. Valdivia, 1965. Effects of simulated high altitude on growth rate of albino guineapigs. *J. appl. Physiol.*, 20 (5): 1022–1025.

65. Delaquerriere-Richardson, L. & E. Valdivia, 1967. Effects of simulated high altitude on pregnancy: Placental morphology in albino guineapigs. *Arch. Physiol.*, 84 (4): 405–417.

66. Dohan, F. C. 1942. Effect of low atmospheric pressure on the adrenals, thymus and testes of rats. *Proc. Soc. exptl. Biol. Med.*, 49: 404–408.

67. Duggar, B. M. 1936. Biological effects of radiation. New York: McGraw-Hill Book Company. 2 vols.

68. Dutton, R. E., V. Chernick, H. Moses, B. Bromberger-Barnea, S. Permutt & R. L. Riley, 1964. Ventilatory response to intermittent inspired carbon dioxide. *J. appl. Physiol.*, 19 (5): 931–936.

69. Dzhailobaev, A. D., M. M. Mirrakhimov & A. Y. Tiilis, 1967. Gaszoobmen u bolnykh s kronichesskim narusheniem krovoobrashcheniya v usloviyakh vysokogorya. *Sb. nauch. Tr. Kirg. Med. Inst.*, 47: 71–75.

70. Eaton, J. W., G. J. Brewer & R. F. Grover, 1969. Role of red cell 2,3-diphosphoglycerate in the adaptation of man to altitude. *J. Lab. clin. Med.*, 73 (4): 603–609.

71. Edwards, C., D. Heath & P. Harris, 1971. The carotid body in emphysema and left ventricular hypertrophy. *J. Pathol.*, 104: 1.

72. Eiglestreiter, H., A. Kubelka & J. Weimann, 1967. Verhalten von Puls und Blutdruck in grossen Höhe bei körperlicher Anstrengung. Beitrag zur Frage der Höhentoleranz unter PersantinWirkung. *Intern. Z. angew. Physiol. Arbeitsphysiol.*, 24 (3): 268–274.

73. Federova, L. D. 1969. K adaptinomu effektu gidrokortizona pri gipoksi. *Dokl. Akad. Nauk SSSR*, 185 (1): 233–235.

74. Fine, B. J. & J. L. Kobrick, 1969. Note on headache, personality and visual performance at altitude. *Percept. Mot. Skills*, 29 (2): 521–522.

75. Fred, H. L., A. M. Schmidt, T. Bates & H. H. Hecht, 1962. Acute pulmonary edema of altitude: clinical and physiological observations. *Circulation*, 25 (6): 929–937.

203

76. Freyer, D. 1964. Decompression sickness at 18,500 ft. A case history, with comment. *Aerosp. Med.*, 35 (5): 479–481.

77. Fried, W., C. J. Johnson & P. Heller, 1970. Observations on regulation of erythropoiesis during prolonged periods of hypoxia. *Bld. J. Hematol.*, 36 (5): 607–616.

78. Frisancho, A. R. 1975. Functional adaptation to high altitude hypoxia. *Science*, 187: 313.

79. Gamboa, R., M. A. Romero & J. Farji, 1970. Thorax resistivity at sea level and at high altitude. *J. appl. Physiol.*, 28 (1): 75–78.

80. Garbash, C., M. E. Mattiessen, P. Heim & I. Lorenzen, 1969. Arteriosclerosis and hypoxia. *J. artecl. Res.*, 9 (3): 283–294.

81. Garvey, M. B., L. H. Dennis, P. K. Hildebrandt & M. E. Conrad, 1969. Hyperbaric erythraemia: Pathology and coagulation studies. *Brit. J. Haematol.*, 17 (3): 275–281.

82. Gill, M. B., J. S. Milledge, L. G. C. E. Pugh & J. B. West, 1962. Alveolar gas composition at 21,000 to 25,000 ft (6400–7620 m). *J. Physiol. London*, 163 (3): 373–371.

83. Gimmerikh, F. I. & M. I. Aginova, 1969. Aktivatory dykhatelnoi funktsii krovi v usloviyakh nizkoi vysokogroya. *Izv. Akad. Nauk Kirg. SSR*, 1: 46–50.

84. Gomori, P. 1960. The regulation of cardiac output in hypoxia. *Acta Med. Acad. Sci. Hungaricae*, 16 (1): 93–98.

85. Gorbunova, N. A. & G. P. Moskaleva, 1969. Vliyanie kratkovremennoi gipoksicheski gipoksii na pokazateli perifericheskoi krovi i prodolzhitelnost Zhizni eritrositov u sobak. *Fiziol. Zhur.* SSSR I. M. Sechenova, 55 (2): 200–206.

86. Gordan, A. S., F. J. Tornetta, S. A. D'Angelo & H. A. Charipper, 1943. Effects of low atmospheric pressure on the activity of the thyroid, reproductive systems and the anterior lobe of the pituitary in the rat. *Endocr.*, 33: 366–383.

87. Graevskii, E. Y. & M. M. Konstantin, 1957. The absence of protection by histotoxic hypoxia against ionizing radiation. *Dokl. Akad. Nauk SSSR*, 114 (1/6): 465–467.

88. Grant, W. C. & W. S. Root, 1952. Fundamental stimulus for erythropoiesis. *Phys. Rev.*, 32: 449.

89. Gray, G. W., A. C. Bryan, R. Frayser, C. S. Housten & I. D. B. Rennie, 1971. Control of acute mountain sickness. *Aerosp. Med.*, 42 (1): 81–84.

90. Grover, R. F. 1963. Basal oxygen uptake of man at high altitude. *J. appl. Physiol.*, 18 (5): 909–912.

91. Hansen, J. E. & W. O. Evans, 1970. A hypothesis regarding the pathophysiology of acute mountain sickness. *Arch Envir. Health*, 21 (5): 666–669.

92. Hansen, J. E., J. A. Vogel, G. P. Stelter & C. F. Consolazio, 1967. Oxygen uptake in man during exhaustive work at sea level and high altitude. *J. appl. Physiol.*, 23 (4): 511–522.

93. Harclerode, J. E., R. T. Houlihan & A. Anthony, 1962. Thyroid function in rats exposed to simulated high altitude. *Summer Meetg. Amer. Soc. Zool.*, Oregon St. Univ., Corvallis, August 1962. *Amer. Zool.*, 2 (3): 413.

94. Hartley, H. 1971. Effects of high altitude environment on the cardiovascular system of man. *J. Amer. med. Assoc.*, 215 (2): 241–244.

95. Heim, L. M. 1965. Spinal cord convulsion in developing rat at altitude (12,470 ft; 3800 m). *Nature*, 207 (4994): 299–300.

96. Heim, L. M. & P. S. Timiras, 1963. Brain maturation at an altitude of 12,000 ft. *47th ann. Meetg. Fed. Amer. Soc. exptl. Biol. Proc.*, 22 (2): 685.

97. Hornbein, T. F. 1962. Evaluation of iron stores as limiting high altitude polycythemia. *J. appl. Physiol.*, 17 (2): 243–245.

98. Hornbein, T. F. & J. W. Severinghaus, 1969. Carotid chemoreceptor response to hypoxia and acidosis in cats living at high altitudes. *J. appl. Physiol.*, 27 (6): 837–839.

99. Hornbein, T. F. & S. C. Sorensen, 1969. Ventilatory response to hypoxia and hypercapnia in cats living at high altitudes. *J. appl. Physiol.*, 27 (6): 834–836.

100. Hultgren, H. N., W. S. Pickard & C. Lopez, 1962. Further studies of the high altitude pulmonary oedema. *Brit. Heart J.*, 24 (1): 95–102.

101. Ikegami, H., Y. Yamazaki, C. Sakakihara, N. Yusa & R. Yurugi, 1969. Studies on the evaluation of physiological load of work at high altitude by means of heart rate. *Rep. Aerosp. Lab.*, 9 (4): 232–241.

102. Ivanov, P. N. 1963. K vaprosu o patogeneze vysotnoi emfizemy. *Patol. Fiziol. Eksptl.*

Terap. 7 (2): 15–19.

103. Jackson, F. 1968. The heart at high altitude. *Brit. Heart J.*, 30 (3): 291–294.

104. Johnson, P. K. & G. A. Feigen, 1962. Growth rate and blood volume in two strains of rat at natural altitude of 12,470 ft. *Stanford Med. Bull.*, 20 (2): 43–55.

105. Kellogg, R. H. 1962. Effect of altitude on respiratory regulation: Regulation and respiration. *Ann. N.Y. Acad. Sci.*, 109 (2): 815–825.

106. Klain, G. J. & J. P. Hannon, 1970. High altitude and protein metabolism in the rat. *Proc. Soc. exptl. Biol. Med.*, 134 (4): 1000–1004.

107. Klausen, K., D. B. Dill & S. M. Horvath, 1970. Exercise at ambient and high oxygen pressure at high altitude and at sea level. *J. appl. Physiol.*, 29 (4): 456–563.

108. Kobrick, J. I. & E. R. Dusek, 1970. Effects of hypoxia on voluntary response time to peripherally located visual stimuli. *J. appl. Physiol.*, 29 (4): 444–448.

109. Kovalenko, E. A. 1961. Oxygen tension in the canine brain under high altitude conditions. *Fiziol Zhur, SSSR* (Transl.) 47 (9): 47–53.

110. Krüger, P. 1929. Die Bedeutung der ultraroten Strahlen für den Wärmeaushalten der Poikilothermen. *Biol. Ztlbl.*, 49: 65–82.

111. Krüger, P. & F. Duspiva, 1933. Die Einflusse der Sonnenstrahlen auf die Lebensvorgänge der Poikilothermen. *Biol. Ge.*, 17: 168–199.

112. Kuhnau, J. 1964. Lebensvorgänge bei tieferen Temperaturen. *Naturw. Rundsahau*, 17 (12): 465–467.

113. Lahiri, A. 1964. Pulmonary ventilation of man at altitude. *Indian J. Pharmacol.*, 8 (2): 31–39.

114. Lauer, N. V. 1961. Do pytannya pro shlyakhy vyvchennya roli vyshahykh viddiliv mozku v reaktsiyi orhanizma na hipoxiyu. *Fiziol Zhur. Akad. Nauk* Ukrain, SSR, 7 (3): 376–384.

115. Lauer, N. V. & A. Z. Kolchynska, 1960. Vplyv detserebratsyi na zimimy dykhannya u krolykiv i holubiv pri hipoksiyi. *Fiziol Zhur. Akad. Nauk* Ukrain, SSR, 6 (4): 490–497.

116. Mani, M. S. 1957. The abominable snowman. *Turtox News*, 35 (8): 3.

117. Mani, M. S. 1968. Ecology and Biogeography of High Altitude Insects. The Hague: Dr W. Junk Publishers, pp. 527.

118. Mani, M. S. 1974. Fundamentals of High Altitude Biology. New Delhi: Oxford & IBH Publishers, pp. 196.

119. Martineaud, J. P., J. Drand, J. Coudert & S. Seroussi, 1969. La circulation cutanee au cours de l'adaptation a l'altitude. *Pflügl. Arch. Eur. J. Physiol.*, 310 (3): 209–213.

120. Meerson, F. Z., M. Y. Maizelis, V. B. Malkin, E. M. Leikina, R. M. Kruglikov & N. A. Popko, 1969. Aktivatsiya sinteza belka i RNK v golovnom mozge kak faktor adaptatsii k vysotnoi gipoksii. *Dokl. Akad. Nauk* SSSR, 187 (3): 697–700.

121. Meerson, F. Z., M. Y. Maizelis, V. B. Malikn, N. A. Novikova, L. Y. Golubeva, E. M. Leikina & G. I. Markovskaya, 1969. Rol sinteza nukleinovykh krilot i belkov v miokarde v adaptasii serdtske k dlitenomi deistviyu vysotnoi gipoksii i ispolzovanie etoi adaptatsii serdechnoi nedostatchnosti. *Dokl. Akad. Nauk* SSSR, 184 (2): 500–503.

122. Menon, N. D. 1965. High altitude pulmonary edema. A clinical study. *New England J. Med.*, 273 (2): 66–73.

123. Michel, C. C. & J. S. Milledge, 1963. Respiratory regulation in man during acclimatization to high altitude. *J. Physiol.*, 168 (3): 631–643.

124. Milledge, J. S. 1963. Electrocardiographic changes at high altitude. *Brit. Heart J.*, 25 (3): 291–298.

125. Mirrakhimov, M. M. 1967. Osnovnoi obmen i ego izmeneniya v usloviyakh vysokogorya. *Sb. Nauch. Tr. Kirg. Med. Inst.*, 47: 85–106.

126. Mirrakhimov, M. M. 1969. Serdechno-sosudistaya sistema v usloviyakh vysokogorya. Leningrad, p. 158.

127. Moffat, D. F., C. Rosse, I. H. Sutherland & J. M. Yoffey, 1961. The effect of hypoxia on guineapig bone marrow. *J. Physiol.*, 157 (2): 48.

128. Moncola, F., A. Carcelen & L. Beteta, 1970. Physical exercise, acid-base balance and adrenal function in newcomers to high altitude. *J. appl. Physiol.*, 28 (2): 151–155.

129. Moncola, F., J. Donarye, L. A. Sobrevilla & R. Guerra-Garcia, 1965. Endocrine studies

at high altitudes II. Adrenal cortical function in sea-level natives exposed to high altitudes (4300 metres) for two weeks. *J. clin. Endocrinol. Metabol.*, 25 (12): 1640–1642.

130. Monge, C. 1951. Syndromes biologiques et cliniques produits par les changements d'altitude. *Bull. schweiz. Akad. Med. Wiss.*, 7 (3/4): 187–200.

131. Monge, C. M., P. M. Mori-Chavez & M. San Martin, 1942. Fiziologia de la reproduction en la espermatogenesis en la altura. *An. Fac. Cien. Med. Lima*, 25: 35–40.

132. Moore, C. & D. Price, 1948. A study at high altitudes of reproduction, growth, sexual maturity and organ weights. *J. exptl. Zool.*, 108: 171–216.

133. Myhre, L. G., D. B. Dill, F. G. Hall & D. K. Brown, 1970. Blood volume changes during three-weeks residence at high altitude. *Clin. Chem.*, 16 (1): 7–14.

134. Mylera, K. C. & P. H. Albrecht, 1970. Hematologic response of mice subjected to continuous hypoxia. *Amer. J. Physiol.*, 218 (4): 1145–1149.

135. Nayak, N. C., S. Roy & T. T. Narayanan, 1964. Pathologic features of altitude sickness. *Amer. J. Pathol.*, 45 (3): 381–391.

136. Nelson, D. & M. W. Burrill, 1944. Repeated exposure to simulated high altitude: estrous cycles and fertility of the white rat. *Fed. Amer. Soc. exptl. Biol.*, 3: 34.

137. Nelson, M. L. & H. H. Srebnik, 1970. Comparison of the reproductive performance of rats at high altitude (3800 m) and at sea level. *Intern. J. Biometeorol.*, 14 (2): 187–193.

138. Nevison, T. O. Jr., J. E. Roberts, W. W. Lackey, R. G. Scherman & K. H. Averill, 1962. 1960–1061 Himalayan scientific and mountaineering expedition. 1 USAF high altitude physiological studies. Meetg. Aerosp. Med. Assoc. 1962. *Aerosp. Med.*, 33 (3): 346.

139. Nimeh, W. & L. G. Villegas, 1968. Altitud y sistema digestivo. *Med. Rev. Mex.*, 49 (1048): 617–619.

140. Novakovsky, S. 1922. The effect of climate on the efficiency of the people of the Russian Far East. *Ecology*, 3 (4): 275–283.

141. Novakovsky, S. 1922. The probable effect of climate of the Russian Far East on human life and activity. *Ecology*, 3 (3): 181–201.

142. Novakovsky, S. 1924. Arctic or Siberian hysteria as a reflex of the geographic environment. *Ecology*, 5 (2): 113–127.

142a. Novikova, N. A. & V. I. Kapelko, 1970. Dinamiko sokratitelnostnoi funkstii miokarda v protsesse adaptatsii k vysotnoi gipoksii i posle ee prekrashchemiya. *Byull. Eksp. Biol. Med.*, 70 (11): 30–32.

143. Numata, M. 1975. Mountaineering of Mt Makalu II and scientific studies in Eastern Nepal 1971. Brief note on an accident. pp. 69–71.

144. Penaloza, D. & F. Simé, 1969. Circulatory dynamics during high altitude pulmonary edema. *Amer. J. Cardiol.*, 23 (3): 369–378.

145. Penna, M., F. Limares & L. Caceres, 1965. Mechanism for cardiac stimulation during hypoxia. *Amer. J. Physiol.*, 208 (6): 1237–1242.

146. Perovic, L., R. Debijadji & V. M. Varagic, 1969. The effect of simulated altitude on the catecholamine content in the brain of rat. *Acta Biol. Jugoslav.*, (C) *Physiol. Pharmacol. Acta*, 6 (2): 215–221.

147. Petrov, I. R. 1955. Eksperimentalnoe izuchenie kislorodnogo gologniya golovnogo mozga i ego znachenie dlya kilink. *Fiziol. Zhur. SSSR*, 41 (1): 9–18.

148. Picón-Réategui, E. 1962. Studies on the metabolism of carbohydrates at high altitudes. *Metab. Clin. & Exptl.*, 11 (11): 1148–1154.

149. Picón-Réategui, E., R. Buskirk & P. T. Baker, 1970. Blood glucose in high altitude natives and during acclimatization to altitude. *J. appl. Physiol.*, 29 (5): 560–563.

150. Piwonka, R. W. & C. D. Barnes, 1970. Spinal reflex activity during acute hypoxia in normal and chronic altitude-exposed cats. *J. appl. Physiol.*, 29 (1): 96–102.

151. Pugh, L. G. C. E. 1964. Animals in high altitudes: Man above 5000 metres, mountain exploration. In: Adaptation to the environment. Baltimore: William & William & Co. Handbook Physiol., 4: 861–868.

152. Rakhimov, Y. A., L. E. Etingen, V. Sh. Belkin & M. M. Usanov, 1969. Morfologiya vnutrennikh organov v usloviayakh vysokogorya. *Tr. Tadz. Med. Inst.*, 10: 1–213.

153. Recavarren, S. & J. Arias-Stella, 1962. Topography of right ventricular hypertrophy in

children native to high altitude. *Amer. J. Pathol.*, 41 (4): 467–475.

154. Reed, D. J. 1960. Effect of sleep on hypoxic stimulation of breathing at sea level and altitude. *J. appl. Physiol.*, 15 (6): 1130–1134.

155. Reeves, J. T., E. B. Grover & R. F. Grover, 1963. Pulmonary circulation and oxygen transport in lambs at altitude. *J. appl. Physiol.*, 18 (3): 560–566.

156. Reeves, J. T., J. Halpin, J. E. Cohn & F. Doud, 1970. Increased alveolar-arterial oxygen difference during simulated high altitude exposure. *J. appl. Physiol.*, 27 (5): 658–661.

157. Riedstra, J. W. 1963. Influence of central and peripheral pCO_2 (pH) on the ventilatory response to hypoxic chemoreceptor stimulation. *Acta Physiol. Pharmacol. Neer*, 12 (4): 407–452.

158. Rizzo, P. J. 1968. Altitude e trbalho fisico. *Rev. Brasil. Med.*, 25 (8): 567–569.

159. Rojas, G. S. & C. Marquez, 1962. Some features of pulmonary insufficiency at high altitudes. *Amer. Rev. Resp. Dis.*, 85 (1): 25–29.

160. Rosenstein, R. & C. M. Moore, 1968. Reversible physiologic disability as a criterion for altitude tolerance in rat. *J. appl. Physiol.*, 24 (5): 733–735.

161. Roy, S. B., J. S. Gulera, P. K. Khanna, J. R. Talwar, S. C. Manchanda, J. N. Pande, V. S. Kausik, P. S. Subba & J. E. Wood, 1968. Immediate circulatory response to high altitude hypoxia in man. *Nature*, 217 (5134): 1177–1178.

162. Saltin, B. 1967. Aerobic and anaerobic work capacity at 2300 metres. *Med. Thorac.*, 24 (4): 205–210.

163. Schnakenberg, D. D. & R. F. Burlington, 1970. Effect of high carbohydrate, protein and fat diets and high altitude on growth and caloric index of rats. *Proc. Soc. exptl. Biol. Med.*, 134 (4): 905–908.

164. Schneider, E. C. 1921. Physiological effects of altitude. *Physiol. Rev.*, 1: 631–659.

165. Sergienko, A. V. 1970. Kompleksoe vliyanie na organism zhivotnykh ostroi gipoksii i vysokoi temperatury oeruhayushchei sredy. *Kosm. Biol. Med.*, 4 (3): 13–18.

166. Simanovskii, L. N. & N. M. Livishits, 1969. Izmeneni glikolita v tkanyakh mozga pri trenirovke k gipoksii. *Vop. Med. Khim*, 15 (1): 66–70.

167. Simhadri, P. 1970. Effect of simulated high altitude on brain glutamic acid. *Indian J. med. Res.*, 58 (7): 908–912.

168. Simhadri, P. 1971. Survival time of mice in simulated high altitude. *Indian J. med. Res.*, 58 (9): 1244–1248.

169. Simhadri, P. & V. V. S. Rao, 1964. Brain glutamic acid at high altitude. *Indian J. Pharmacol.*, 8 (2): 10.

170. Simonson, E. 1961. Effect of age on changes of extra-cranial circulation during hypoxia. *Circulation Res.*, 9 (1): 18–22.

171. Simmonds, D. H. & E. H. Kahn, 1966. Blood gases of rats at altitude and sea level. Proc. Intern. Symp. Metab. adapt. to temperature and altitude. Kyoto (Japan) Sept. 13–17 1965. *Proc. Fed.*, 25 (4.1): 1247–1252.

172. Slater, J. D. H., E. S. Williams, R. H. T. Edwards, R. P. Ekins, P. H. Sonksen, C. H. Beresford & McLaughlin, 1969. Potassium retention during the respiratory alkalosis of mild hypoxia in man. Its relationship to aldosterone secretion and other metabolic changes. *Clin. Sci. London*, 37 (2): 311–362.

173. Slonim, A. D. 1957. Znacheni prirodnykj faktorov vneshnei sredy dlya fiziologii selskokhozyaistvennykh zhivotnykh. Voprosy fiziologii selskokhzyaistvnnykh zhivotnykh. Akad. Nauk SSSR, 173–179. *Zhur. Biol.*, 1959 54079.

174. Smith, E. E. and J. W. Crowell, 1963. Influence of hematocrit ratio on survival of unacclimatized dogs at simulated high altitude. *Amer. J. Physiol.* 205 (6): 1172–1174.

175. Sobrevilla, L. A., I. Romero, F. Kruger and J. Whittembury, 1968. Low oestrogen excretion during pregnancy at high altitude. *Amer. J. Obst. Gynaecol.*, 102 (2) (6): 828–833.

176. Sobrevilla, L. A. and F. Salazar, 1968. High altitude hyperurecemia. *Proc. Soc. exptl. Biol. Med.*, 129 (3): 890–895.

177. Sorensen, S. C., 1970. Ventilatory acclimatization to hypoxia in rabbits after denervation of peripheral chemorecepters. *J. appl. Physiol.*, 28 (6): 836–839.

207

178. Stampfer, D. A., R. A. Kinsman and W. O. Evans, 1970. Subjective symptomatology and cognitive performance at high altitude. *Percept. Mot. Skill,* 31 (1): 247–261.
179. Stenberg, J., B. Ekblom and R. Messin, 1966. Hemodynamic response to work at simulated altitude 4000 m. *J. appl. Physiol.,* 21 (5): 1489–1595.
180. Strickland, E. H., E. Ackerman and A. Anthony, 1962. Respiration and phosphorylation in liver and heart mitochondria from altitude-exposed rats. *J. appl. Physiol.,* 17 (3): 535–538.
181. Sundstroem, E. S. and G. Michaels, 1942. The adrenal cortex in adaptation to altitude, climate and cancer. *Mem. Univ. California,* 12: 1–410.
182. Syrotynin, N. N. 1942. Zhittya navisotak i kovorova visoti.
183. Syrotynin, N. N. 1964. Vliyanie adaptstsii k gipoksii i akklimatizatsii k vysokogornomu klimatu na ustoichivost zhivotnykh k nekortym ekstremalnym vozdeistviyam. *Patol. Fiziol. Eksp. Terap.,* 8 (5): 12–15.
184. Syrotynin, N. N. 1965. Pro rizni varanty akklimatyzatsiyi do vysokhirnoho klimatu. *Fiziol. Zhur. Akad. Nauk. UKRSSR,* 11 (3): 283–288.
185. Syrotynin, N. N. 1969. Do evolyutsiyi dykhalnoi funktsyi. *Fiziol. Zhur. Akad. Nauk UKRSSR,* 15 (2): 193–199.
186. Syrotynin, N. N. 1970. Deyaki pidsumky vyvchenya chervonoyi krovi na hirskykh visotokh. *Fiziol. Zhur. Akad. Nauk UKRSSR,* 16 (2): 205–210.
187. Terjung, W. H. 1970. The energy budget of man at high altitude. *Intern. J. Biometeor.,* 14 (1): 13–43.
188. Terman, J. W. and J. L. Newton, 1964. Changes in alveolar and arterial gas tensions as related to altitude and age. *J. appl. Physiol.,* 19 (1): 21–24.
189. Thomas, R. E. and J. E. Kittrell, 1966. Effect of altitude and season on the canine hemogram. *J. Amer. vet. Med. Assoc.,* 148 (10): 1163–1167.
190. Tiisala, R., 1962. Endocrine response to hyperoxia and hypoxia in the adult and newborn rat. An experimental study with radio-active phosphorus. *Ann. Acad. Sci. Fennicae,* (A. V. Med.) 95: 5–141.
191. Tribukait, B. 1963. Über den initialen Ansteig der O_2 Kapazität des Blutes den Ratte bei Hypoxie. *Acta Physiol, Scand.,* 57 (1/2): 90–98.
192. Tribukait, B. 1963. Der Einfluss chronischer Hypoxia entsperechend 1000–8000 m Hohe auf die Erythropoiese der Ratte. *Acta Physiol. Scand.,* 57 (1/2): 1–25.
193. Tsagareli, Z. G. & G. I. Natsvilishvili, 1969. Nekotore ultrastrukturnye pokazateli sostoyaniya miokardialnykh kletok pri obschei gipoksii organizma. *Soobshch. Akad. Nauk Gruz. SSR,* 54 (1): 233–236.
194. Tukhtaev, T. M. & S. I. Pauk, 1970. Izmenenie sodezzheniya obschego belika i belkovykh fraktsii v syvorotke krovi krys v usloviyakh vysokogorya. *Kosm. Biol. Med.,* 4 (1): 81–82.
195. Vasilenko, M. E., O. G. Gazenko, P. M. Graminitskii, A. G. Zhironkin, V. N. Zvorykin and A. G. Kuznetsov, 1958. Izmeneniya vysotnoi usloichivosti pri barokameronoi trenirovke: Funktsii organizma v unloviyakh izmeneniya gavozoi sredy. Akad, Nauk SSSR, 2: 137–152.
196. Vinokurov, B. A., A. E. Gerov, V. P. Kurkovskii & I. D. Karchenko, 1958. K vaprosu o funktionalnykh i morfologicheskikh izmeneniyakh tsentralnoi nervoi sistemy pri raz-rezhnii atmosphery sootvetstuyuschem vysote 18,000 metrov. Funktsii organizma v suloviyakh izmeniya gazovoi sredy. Akad. Nauk SSSR, 2: 66–80.
197. Visvanathan, R., S. K. Jain, S. S. Subramanian, T. A. V. Subramanian, G. L. Dua & J. Giri, 1969. Pulmonary edema of high altitude. *Amer. Rev. Resp. Dis.* 100 (3): 334–341; 342–349.
198. Vogel, J. A., G. W. Bishop, R. I. Genovese & T. L. Powell, 1968. Hematology, blood volume and oxygen transport of dogs exposed to high altitude (aerospace). *J. appl. Physiol.,* 24 (2): 203–210.
199. Vogel, J. H. K. 1970. Advances in cardiology. Vol. 5. Hypoxia, high altitude and heart. Symposium, viii-195. Basel: S. Karger.
200. Ward, M. 1975. Mountain Medicine. A Clinical Study of Cold and High Altitude. London: Crosby Lockwood Staples.

201. Weihe, W. H. 1967. Effects of ambient air temperature on the acclimatization of rats to high altitude. Proc. 3rd intern. Biometeor. Conger. 1963. *Biometeor.*, 2 (1): 219–225.

202. Weil, J. V., E. Cyrne-Quinn, I. E. Sodal, W. O. Friesen, B. Underhill, G. F. Filley & R. F. Grover, 1970. Hypoxic ventilatory drive in normal man. *J. clin. Invest.*, 49 (6): 1061–1072.

203. West, J. B. 1963. Work limitations at very high altitude. *J. Physiol.*, 167 (1): 14.

204. West, J. B., 1967. Exercise limitations at increased altitudes. *Med. Thorac.*, 24 (6): 233–337.

205. Whittembury, J., R. Lozano, C. Torres & C. Monge, 1968. Blood viscosity in high altitude polycythemia. *Acta Physiol. Latinoamer.*, 18 (4): 355–359.

206. Whitten, B. K., R. F. Burlington, M. A. Posiviata, C. M. Sidel & G. R. Beecher, 1970. Amino acid catabolism in environmental extremes: Effect of high altitude and calories. *Amer. J. Physiol.*, 218 (5): 1346–1350.

207. Wilcox, B., W. C. Austen & H. W. Bender, 1964. Effect of hypoxia on pulmonary artery pressure of dogs. *Amer. J. Physiol.*, 207 (6): 1314–1318.

208. Will, D. H., A. F. Alexander, J. T. Reeves & R. F. Grove, 1962. High altitude induced pulmonary hypertension in normal cattle. *Circ. Res.*, 10 (2): 172–177.

209. Wollaston, A. F. R. 1922. Natural History. In: Mount Everest. By C. K. Howard-Bury. London: Edward Arnold & Co. p. 356.

210. Wood, J. E. 1970. The relationship of peripheral venomotor response to high altitude pulmonary edema in man. *Amer. J. med. Sci.*, 259 (1): 56–65.

211. Wooley, D. E. & P. S. Timiras, 1963. Changes in brain glycogen concentration in rats during high altitude (12,470 ft) exposure. *Proc. Soc. Exptl. Biol. Med.*, 114 (3): 571–574.

212. Zimmer, K. G. 1961. Studies on quantitative radiation biology. Translation by H. D. Griffith. Edinburgh: Oliver & Boyd.

V Man in highland ecosystem: human acclimatization to highland conditions

M. S. Mani

1. Introduction

Although the contrary was firmly believed in the early days of mountaineering, especially in the region of Mt Everest, it soon became apparent that acclimatization to the extremes of high altitude conditions is possible in man. The problem of human acclimatization has since then been rather intensively studied in the field by mountaineering expeditions and in the laboratory under experimental conditions. Yet it cannot be said even today that the basic features are correctly understood. The difficulty lies partly in the erroneous approach to the problem and partly in prevailing prejudices in biology.

Considerable confusion seems to prevail at present about the precise meanings and limits of the expressions *adaptation* and *acclimatization*. Monge (28) considers, for example, that adaptation is purely temporary, but acclimatization is permanent. He thus observes 'when adaptation is over, acclimatization supervenes'. The Russian workers have added their own contributions to the prevailing confusion. They make a wholly unnecessary difference between adaptation and acclimatization by defining adaptation as a process stimulated by hypoxia and acclimatization as a genetic process. All this confusion stems naturally from the erroneous darwinian concept of 'selection' which has retarded progress. It must be stressed here that acclimatization is merely one phase of adaptation to a specific environment, determined largely by *atmospheric* conditions – *climate alone*.

Adaptation in man is basically the same as in other animals, but unlike in most other animals, in man it represents the sum total *not only of his physiological* but also psychological and only to a very minor degree visible structural adjustments, when exposed to unusual environmental stresses, in order to restore and maintain homeostasis. Adaptation particularly to climatic conditions is generally spoken of as acclimatization. As may be understood, the ability to acclimatize differs greatly in different individuals. It may be temporary or long-standing (permanent), incomplete or 'complete', but fundamentally it is a part of the *reaction to specific climatic factors*.

2. Basic Features

Monge (28) and Monge (29) gave excellent accounts of the acclimatization of the residents of high altitude of the Peruvian Andes. Though individuals can acclimatize to a certain extent within their life time, true acclimatization,

211

recognized by unimpaired reproductive ability, requires continuous residence and reproduction for many generations. Some workers like Kaulbersz (20) consider that the best index of high altitude acclimatization in man is his tolerance to maximum physical exercise. Acclimatization to high altitude may generally be recognized by haematogenic function. When acclimatization has been achieved, the activity of the heart is economical, suggesting a lowered requirement of oxygen for the same given supply of energy. Acclimatization is also generally associated with less marked intensification of external respiration, compared to the considerable alveolar ventilation. The anaerobic productivity of habitual athletes utilizes forms of energy arising from intensity of oxygenation of the cellular system, utilization of anaerobic resynthesis of ATP, etc. Oxygen consumption is normal in cellular level in the acclimatized individual. Structural changes that arise in course of acclimatization are determined primarily in macromolecular level. The cellular process involves the basic process of metabolism; this is extremely important in the acclimatization syndrome.

Like all reactions of organisms to external changes, acclimatization to high altitude commences in man after a more or less brief latent period on ascent of high mountain. The various reactions and changes described in earlier chapters as effects of exposure to high altitude hypoxia must really be considered as representing in essence the various phases of adjustments to the altered conditions. These reactions tend to bring about a gradual acclimatization of the sea-level resident to the high altitude environment.

Some of the specific and non-specific lines of acclimatization of man to high altitude have been discussed in detail by Barbashova (6). The dynamics of the adaptive reactions at cellular levels during training of experimental rats to high altitude hypoxia, investigated by the same author (7) and her investigations on the mechanism of resistance of hypoxia have considerable relevance to human acclimatization to high altitude (7). Different types of acclimatization and the resistance of acclimatized animals to extreme factors of high altitude were studied by a number of Russian workers like Syrotynin (38). The changes in the basal metabolism and respiration in the course of high altitude acclimatization of man are described by Gill & Pugh (13). Miles (26) has suggested that the ability of Hindu saints to exist at high altitude on the Himalaya is to be attributed to their respiratory training or more correctly controlled breathing or regulated breathing as part of *hata yoga* practice, which has much in common with high altitude adaptation. The practice of regulated breathing serves to fortify the individual against an early onset of hypoxic symptoms.

The problem of acclimatization of athletes to high altitude has been studied by a number of workers in South America and in Russia (31, 32). The limitations of physical work at very high altitude in relation to acclimatization, discussed by West (40, 41) are interesting in this connection. While some workers believe that physical exercise favours rapid acclimatization of man to high altitude, others think exercise has no special value more than that at sea level in improving the respiratory efficiency (12, 14). The relations between acid-base balance, adrenal function, physical exercise and

212

acclimatization to high altitude are discussed by Moncola (27). The changes in gas exchange when completing exercise during high altitude ascent was studied by Agadzhanyan *et al.* (2). The respiration of man during physical exercise at high altitude during acclimatization is described by Lahiri *et al.* (22) Exercise at ambient and high oxygen pressure at high altitude during acclimatization is dealt with by Klause *et al.* (21). Other important contributions on acclimatization are listed in the bibliography (1, 4, 9, 32, 33, 34, 35, 36, 37).

In the course of the acclimatization reaction, the increasing deficiency of oxygen of the air and the fall in the atmospheric pressure with increase in altitude induce specific reactions in the organism so as to counteract the deleterious effects of these changes. Broadly speaking, we may observe two phases of this process, viz. (i) adaptation to hypoxia and (ii) increase in supply of oxygen. These reactions commence in subcellular, cellular and tissue levels and gradually extend to major function organ systems. There is more or less pronounced intensification of anoxibiotic processes, anaerobiosis, capacity for maintaining the internal tissue oxidation under inadequate supply of oxygen, limiting the need for oxygen, re-utilization of enzymatic reactions of cellular oxidation-reduction, etc. The reaction favours bradicardia, vagotony, fall in gas exchange, reduction of general functional activity and significant storing up of reserve energy. The retardation of flow for capillary vasodilation and dilatation of the pulmonary circulation, which were mentioned in earlier chapters, are parts of this reaction. The cardiac hypertrophy and gradual rise of venous pressure must also be considered as parts of the same phenomena.

The increase in haematocrit and haemoglobin content of peripheral blood during ascent of high altitude, the predominance of erythroblasts over myeloids and marked rise in erythropoiesis are also adaptive reactions of a major nature.

The lowered thyroid function has the important chain effect of reducing the cortical sensitiveness, thus permitting an increased tolerance of hypoxia; it also compensates hypoxia. Anaerobic glucolytic catabolism then satisfies the requirements of energy for the maintenance of muscular activity to a large extent. All aerobic functions then become diminished and at the same time anaerobic reactions increase. The sensitivity of the heart and the blood vessels is diminished and there is also a corresponding rise of endocrinal factors. Insufficient myocardia increases the catecholamines with excess energy. It is really the changes in the chemistry of the tissues, reacting to hypoxia, that underlie the physiological reactions of man, expressed as metabolic processes and exhibited in various conditions, which produce 'stress', resistance to ionizing radiations, etc. This involves essentially a reorganization of the physico-chemical properties at cellular level and an intensification of external respiration by increase in haemoglobin, ascorbic acid, glutathion, cytochrome-C, cytochrome oxidase, succinic oxidase, ATP, etc.

The adaptation to high altitude hypoxia may thus be described as consisting in the restriction of the oxygen consumption, of the activity of certain oxidative

213

enzymes and corresponding increase of anaerobic glycolysis and glycogensis.
The ATP intensifies its activity in tissue and the increase in metabolism in its
enzyme system and the stimulation of anaerobic processes raises the resistance with respect to altitude stress. Acclimatization is effected mainly by
diminishing the pressure gradient of oxygen from ambient air to its site of
utilization at mitochondria. Hyperventilation is the principal factor in reducing the oxygen cascade. This increases the partial pressure of oxygen in the
alveolar space. Hyperventilation is initially due to stimulation of peripheral
chemoreceptors, leading to respiratory alkalosis. The alkalosis is compensated gradually by renal extraction of excess bicarbonate and a more rapid
compensation by changes in cerebrospinal fluid. Hypoxia and increased
sensitivity of the respiratory centre to CO_2 serve to sustain hyperventilation.
The transport of oxygen to tissue is basically a function of cardiac output,
the oxygen content of the blood in the systemic circulation and affinity of
haemoglobin for oxygen. The erythrocyte count and haematocrit are raised.
The haemoglobin level increases to 20 gram per 100 cc or even higher. The
haemoglobin-oxygen dissociation is characterized by a relatively high PO_2 in
the capillaries to facilitate diffusion of oxygen to the tissues. Increased
2,3-diphosphoglycerate enters the core of the haemoglobin molecule to
stabilize the deoxy-form and to release oxygen. Increased capillary density
reduces the distance of oxygen diffusion and increased myoglobin concentration facilitates the rapid oxygen diffusion to tissues.

To summarize, the process of acclimatization of the sea-level resident to
high altitude thus involves two fundamental changes (Fig. 14-1).

1. *Adjustment of hypoxia:* This occurs mainly, if not exclusively, in cellular-
tissue level in three principal phases: (i) fall in consumption of oxygen (ii)
intensification of anaerobic glycolysis and (iii) rise of resistance of tissue to
deficient oxygen.
2. *Increase in oxygen supply:* This occurs partly in cellular-tissue level and
largely in functional systems. The cellular-tissue level process comprises (i)
increase of myoglobin (ii) changes in enzymatic activity, and (iii) increase in
utilization of the oxygen from PO_2. The processes in functional system
include (i) hyperventilation (ii) rise in blood flow (iii) polycythemia (iv)
changes in enzyme activity and (v) changes in acid-base equilibrium.

3. Factors Influencing Acclimatization

While acclimatization is directed primarily to readjustments that would
enable man to live under the conditions of low oxygen content of air at high
altitude, the process is more or less profoundly influenced by number of
other environmental factors operating at high altitude. The acclimatization
of man to high altitude is in reality a complex process, which is influenced by
the following important factors (Fig. 14-2): (i) individual factors (ii) previous
experience of high altitude (iii) rate of ascent of high altitude (iv) the altitude

214

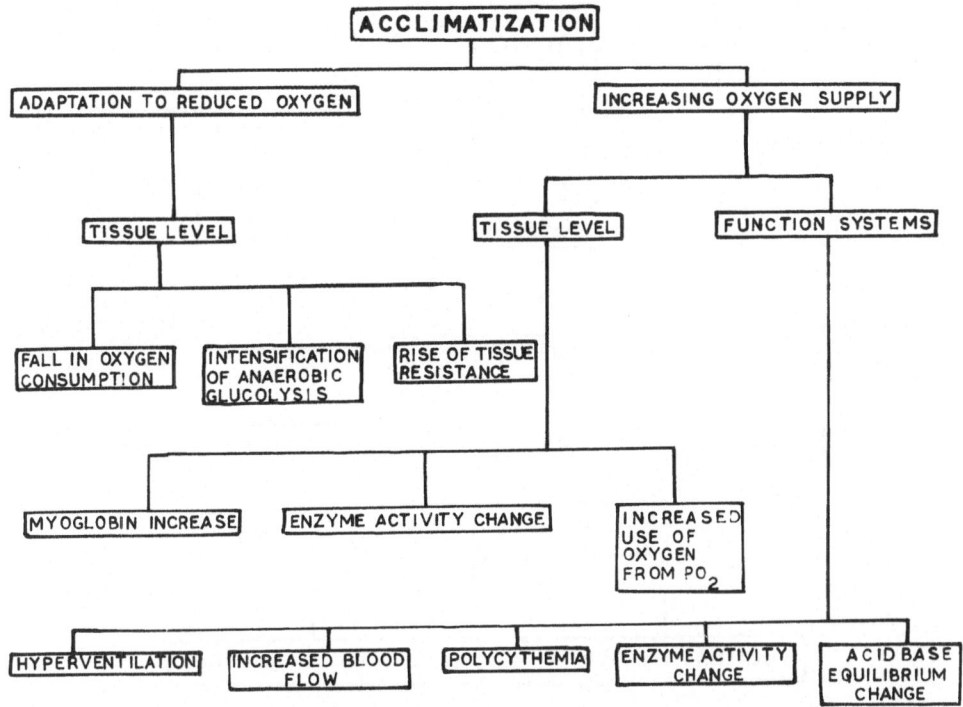

Fig. 14-1. The fundamental processes involved in the acclimatization of the sea-level resident to the high altitude environment, which bring about partial adjustment of the body organization and functions to the decreasing supply of oxygen and partial rise in supply of oxygen, both in the tissue and function levels.

(v) the duration of residence at high altitude (vi) activity at high altitude and (viii) the mountain constant.

i. *Individual Factors*

Although precise observational data are not available, it is nevertheless evident from our experience on the Himalaya that the intensity and facility of acclimatization to high altitude differ in different individuals. The influence of age on the effects of hypoxia has already been mentioned. Closely bound up with differences in age are the physiological and nutritional conditions and the habits of the individual. Under conditions of abnormal physiology and nutritional deficiency, the acclimatization of an individual to high altitude may not be expected to be either complete or early; there is usually considerable retardation. Our experience on the Himalaya and the observations of workers in other parts of the world have shown that acclimatization sets in much earlier and is more nearly complete in individuals, who habitually engage themselves in active physical habits while at sea level, than in individuals of sendentary habits. Field biologists did generally better with us in our expeditions to the Himalaya than hardened museum

215

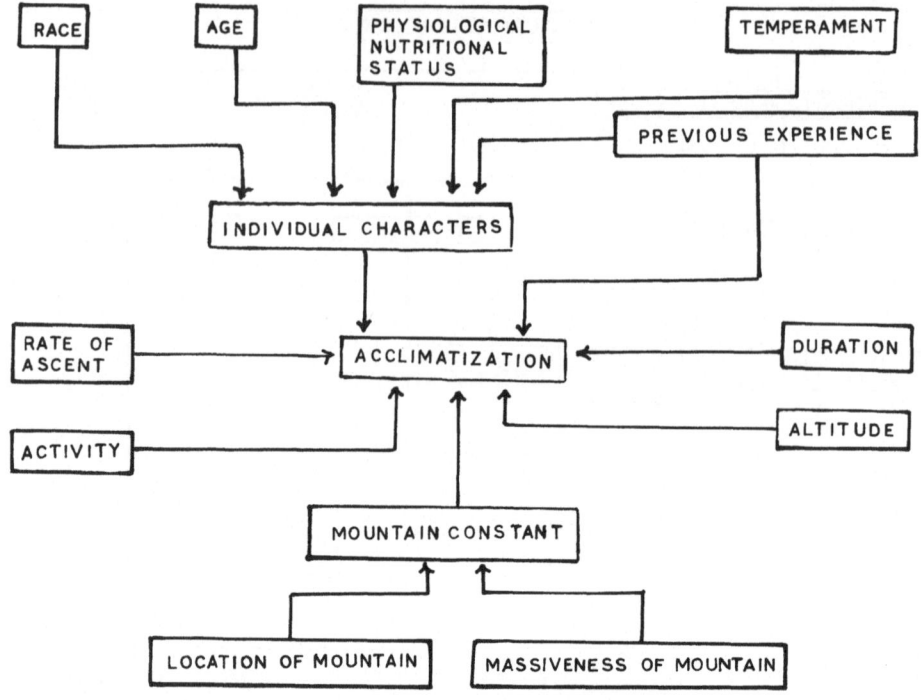

Fig. 14-2. Some of the principal factors which influence the acclimatization of the sea-level resident to the high altitude environment.

workers. There is also some evidence to believe that individuals, who are permanent residents of the mountain valleys, develop high altitude acclimatization more readily than inhabitants of the far-off plains. This is largely in consonance with our experience with porters engaged from the valleys of Nepal, Kulu, Lahaul and Spiti in our entomological expeditions to the Himalaya, whose performance proved on the whole better than of those brought up from the plains of India.

Given all the conditions that apparently favour rapid and satisfactory acclimatization, it has been repeatedly observed that unlike in the case of various experimental animals, the acclimatization process in man is profoundly influenced by psychological factors of the individual, such as his mental attitude and temperament. The temperamental conditions often completely mask the influence of all other factors and either inhibit, retard or accelerate the acclimatization process. Hunt (10) remarks, for example, that the capacity to ascend to elevations of 6000 m on the Himalaya varies greatly between individuals; some men do not seem to be able to go above 6450 m and probably only exceptional individuals can go above 8230 m without the use of oxygen. He also found wide variations in this capacity in the same individuals from day to day. This fact would explain, at least partly, the paradox of individuals, physically fit and strong, with nearly all the metabolic processes normal and indeed apparently wholly conducive to a high degree of acclimatization and even showing certain necessary preliminary reactions like increased haematogensis and hyperventilation, continuing to suffer from acute mountain sickness, after others have nearly all recovered. The only explanation in such cases seems that they *do not want* to be acclimatized; their mental attitude seems to serve as a *built-in physiological block* in some unexplained manner, perhaps through neurohormonal mechanism. In man at least this is perhaps the most important factor, which very profoundly influences the acclimatization process; all other factors would seem to be of minor importance. The *actue anxiety to get back quickly to the habitual mode of life effectively retards or completely inhibits acclimatization.*

We may also draw attention to the differences in degree of acclimatization that must be traced to differences in race, which are in turn determined by marked differences in stature, physical capacity and other characters.

ii. *Previous Experience of High Altitude*

It is well known that when an organism reacts to some specific factor, the reaction disappears when the factor ceases to operate. The reaction leaves, however, the organism in such a state that if the same factor operates a second time, within a reasonable interval, the reaction appears more readily than on the first occasion and against a lesser intensity of the factor than earlier. Each successive action of the factor leaves an invisible impression on the organism, which thus eventually comes to be attuned to the specific factor. This attunement capacity is well known as *facilitation* in physiology and explains the relation between previous experience of high altitude conditions and the degree of acclimatization and the readiness with which it arises in man, who has been on the mountain before. We have empirical field evidence from the Himalaya in support of this conclusion. The acclimatization, which develops in a man ascending a high mountain, naturally disappears on his return to the plains, yet it leaves him 'attuned' to the high altitude, so that during his next ascent of a high mountain, he suffers much less severely from high altitude sickness and becomes also acclimatized sooner than on the previous occasion.

iii. *The Rate of Ascent of High Altitude*

Rapid and abrupt change from the sea level to high altitude conditions usually induces violent symptoms and causes profound disturbances in physiology. Such an ascent leaves inadequate time to the tissue to readjust to the all-too-sudden deprivation of oxygen supply. The acclimatization is, therefore, either greatly retarded or may even be incomplete and in extreme cases totally fail to arise. The abrupt disruption of normal metabolic processes results in secondary side effects as chain reactions that seem to prevent effective adjustments to the oxygen deficiency. Gradual ascent by stages of a mountain, allowing ample time for the tissues and organ systems to readjust their functions, causes on the other hand much less acute

217

symptoms of high altitude sickness. The acclimatization process that sets in almost as the ascent commences is, therefore, effected earlier and is also more complete than in the case of rapid ascent. It is thus the common experience of mountaineers to rest necessarily slightly at lesser altitude for some days, before undertaking the final ascent to the summit of the mountain.

iv. *Altitude Limits*

Acclimatization is determined largely by the altitude to which man reaches. The human organization functions almost naturally and normally from sea level to certain altitude on mountains, above which the symptoms of hypoxia begin to show. This lower limit of altitude depends on the general climate, the permanent place of residence and other peculiarities of the individual. The mean value of the lower limit may, however, be taken as about 1800 m (Fig. 14-3). Below this altitude, acclimatization is not needed in man nor does it develop; this represents the altitude valence of the human species for optimal conditions of life. Above this limit, the degree of acclimatization progressively rises with increase in altitude up to a maximum limit, between 3600 and 4200 m. Acclimatization becomes less pronounced above this upper limit, falls at first gradually as the altitude increases, but more rapidly about 5000 m and finally there is complete failure to become acclimatized above the critical limit of 6000 m. It is interesting in this connection to refer to the work of Campbell (10), who has shown that mammals cannot acclimatize to altitude of 6000 m or to oxygen content of 10%. It is also interesting to recall here that it is within this region of complete and maximum acclimatization that all the permanent high altitude habitations of

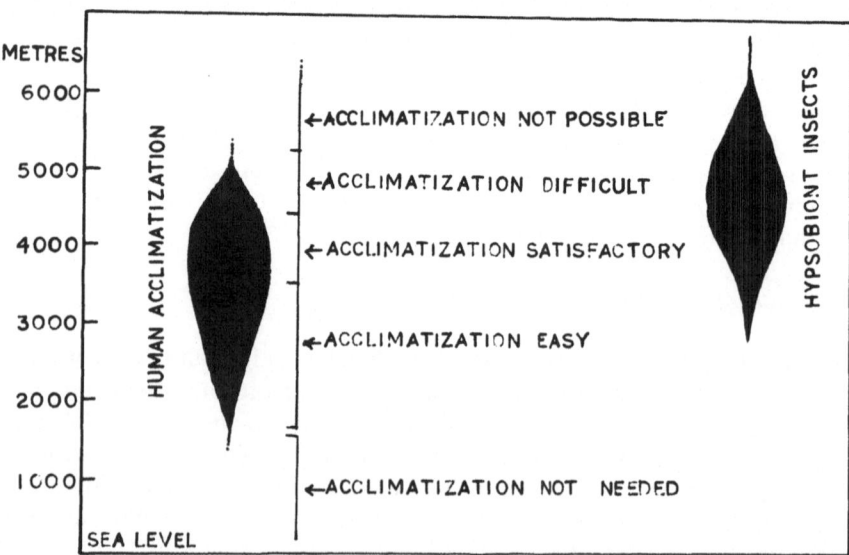

Fig. 14-3. Altitudinal limits of human acclimatization to the high altitude environment. The limits of human acclimatization also roughly coincide with the region of maximum stable hypsobiont communities.

218

human beings in Tibet and the Bolivian and Peruvian Andes happen to be situated. Both below and above these limits of 3600–4200 m, acclimatization disappears or fails to be permanent and satisfactory. It seems, therefore, that human body is wholly incapable of permanent adjustment to the altitude conditions above these limits. High altitude regions above 5000 m must be considered to lie outside of the optimal ecological limits of adjustability. At these altitudes even Tibetans and the Andean inhabitants show acute symptoms of metabolic and other disturbances. The experience of various Mt Everest expeditions has also shown that almost up to 6000 m the use of oxygen is not absolutely necessary. Although man can reach up to altitudes above 8000 m it would be impossible for him to exist permanently in the regions far above 5000 m, because acclimatization fails completely.

The above considerations are of great interest in relation to our observation that the maximum stable hypsobiont insect communities on the Himalaya occur at elevations of about 4200 m (24). It is also at this altitude region that the largest numbers of true high altitude insects occur on the Himalaya; their abundance diminishes both above and below this zone (23, 24).

The acclimatization of man to high altitude is total and stable only within these limits and partial and temporary below 3600 m and above 4200 m. Below 1800–2000 m he does not need acclimatization and above 6000 m there is total absence of acclimatization. Most complete and stable acclimatization is met with at about 4200 m.

v. *Duration of Stay at High Altitude*

As human acclimatization is a complex process, it must be obvious that it cannot be an instantaneous process. The reactions no doubt set in after a latent interval of exposure to high altitude conditions and continue as long as the high altitude conditions prevail and recovery also takes place only some time after return to the sea level. The acclimatization is achieved only after a variable period of residence at high altitude. If the person returns to sea level before this period, acclimatization is naturally very incomplete. This condition has been erroneously described by some authors as adaptation. While temporary acclimatization may arise within two to six weeks of residence at high altitudes between 3000 and 4200 m, total and stable acclimatization often requires two to three years and in some cases even five years uninterrupted residence on mountains. The time required for satisfactory acclimatization is necessarily short at lower elevations and longer at higher elevations. In other words, man requires a much longer period for acclimatization at higher altitudes than lower. The longer the residence on mountains, the more complete and more stable is his acclimatization.

vi. *Activity at High Altitude During Acclimatization*

High altitude acclimatization seems, according to the limited evidence from the Himalaya and judging from the data available in literature, to be favoured by moderate physical activity on the mountain. Violent physical exercise, however, greatly disturbs and retards the acclimatization process. The effect of physical activity is apparently indirect through maintenance of the general tonus of the muscles and other tissues. Perhaps by lowering the general physical fitness, inactivity would appear to adversely influence the development of satisfactory acclimatization.

As already indicated, a great deal of attention has been paid to the problems of physical training of athletes on mountains and the rôle it plays in their high altitude acclimatization (2, 3, 4, 6, 11, 12, 14, 30, 31). Some observers report that vigorous work brings about increased polycythemia, pulse, haemoglobinemia, respiratory rhythm and perceptible capillary permeability. The problem of relation between physical activity and high altitude acclimatization is of great importance in connection with military operations on mountains and much remains yet to be elucidated.

vii. *Mountain Constant*

Even our limited experience on the Himalaya has indicated that human acclimatization to high altitude conditions proceeds wholly differently in different parts of the vast Himalayan system. Other things being assumed to be more or less equal, the ability of man to tolerate the high altitude conditions differs significantly, even at comparable altitudes, in the Eastern, Central and Northwest Himalaya.

Independently of the various factors discussed above, the degree and rate of acclimatization is influenced by the mountain itself. This influence rests perhaps on the differences in latitude, greater continentality, relatively scanty snow cover, high atmospheric aridity, etc., between the eastern and extreme western ends of the Himalaya. It also seems evident that the acclimatization process must follow entirely different courses on single mountains and on massive ranges. In earlier publications on high altitude entomology, we have emphasized the striking differences in the high altitude ecologies of single massifs and large mountain ranges (24). The effects of the massiveness of the mountain show as perceptible retardation of the acclimatization. Man acclimatizes distinctly more slowly on massive mountains like the Himalaya than on single massifs like Mt Kilimanjaro.

The muscle fatigue, the prolonged oxygen hunger of the tissues and diverse other symptoms tend to be rather very much more acute on the Himalaya than on smaller mountain systems. It would seem that it is not simply the amount of oxygen deficiency that underlies this striking difference, because at comparable elevations, ignoring the small differences due to latitude and barometric pressures, the partial pressure of oxygen in the atmosphere may be taken as approximately the same. The correct explanation of this peculiarity must, however, be sought for in the phenomena of gravity anomaly.

It may be useful to recall here that the human skeleton and all his skeletal muscles and diverse other internal organs have evolved under the gravitational force prevailing at or near sea level. They are, therefore, apparently unsuited to function with equal mechanical efficiency under conditions of even slightly higher or lower value of g. They do not thus seem to respond at once. It seems that the rôle of oxygen deficiency of the atmosphere in the onset of rapid muscle fatigue and general exhaustion of man on slight physical exertion on high mountains has so far been consistently exaggerated. While certainly the lowered oxygen content does affect these conditions, it does *not* evidently account for the whole of the difference. Muscle weakness, polycythemia, headache, dizziness and the other symptoms of high altitude sickness, at present supposed to be entirely the result of hypoxia, may in part be the result of the human muscles having to function under distinctly, though small, different values of g on mountains.

Unfortunately this aspect of the relation between gravity and organisms, their movements and functions, has been totally ignored by biologists so far. Though we know of an isolated observation by Korzhuev that the skeleton in the Middle Asiatic oxen constitutes about 17% of the body and the medulla osea about 45% of the skeleton, we have no other precise observational data to enable us to discuss in detail the relation between gravity and the skeleton-muscle complex. It cannot yet be denied that such a relation does exist.

On massive mountains like the Himalaya, man is acted on not only by g but he is also reacting to the added attraction pull due to the massive mountain itself. Perhaps unlike in the case of various other organisms, the close relation between such slight differences in gravity and the functioning of the skeleton-muscle complex is the direct consequence of *his erect posture*. The slight differences in gravity that would perhaps not seriously affect a quadruped, adversely interfere with the normal work load on the human heart by virtue of his erect posture, increases the strain on the heart and decreases the blood flow rate and pressure to head, far more than in a quadruped. The predominantly short stature of the typical high altitude residents nearly all over the world, while partly the result of the thyroid hypo-functioning, may also be correlated with the lowering of the centre of gravity, so that the short stature will have to be considered as a structural adaptation of man for life at high altitude.

4. Epilogue

For almost all forms of life of the lowland biota the high altitude is inhospitable in the extreme, for many of the lowland animals it proves to be a veritable graveyard. Yet we have seen that the high altitude region is alive; a variety of organisms exist in complete harmony with these very unfavourable conditions. An astonishing group of plants and animals have apparently succeeded in adjusting their vital functions so as to enable them to exist at high altitude. How has this been possible for them, when particularly man, with his unique capacity to shape his own special environment, seems to have completely failed?

In classical biology it is the general fashion to speak of the origin and evolution of individual organisms in total isolation from others and from their habitat. We believe, however, that evolution of an organism is only an infinitesimally small phase of the evolution of its ecosystem as a whole (25). The *origin and evolution of the high altitude organisms are integral parts of the evolution of the high altitude ecosystem* – of the birth and rise of the mountain itself. The high altitude plants and animals, now found on the mountain, are in harmony with the high altitude environment not only because they are essential parts of that environment, *but also because they formed an integral part of the very process that shaped the same environment. The same process brought them and the mountain into existence at the same time.*

How can we attempt to explain intelligently the origin and evolution of an ectoparasitic group like, for example, the Mallophaga, without at the same time referring to the host animal? The host-parasite form together an indivisible unit that evolves as on single organism. Likewise the high altitude organisms and the mountain habitat form together an indivisible whole, the high altitude ecosystem that evolves as one.

Unlike the evolution *in situ* of the high altitude organisms, human evolution was not isochronic with the uplift of the mountain systems, on some of which he may be resident today. As is well known, nearly all the important present-day high mountain ranges of the world are products of the Tertiary orogenic activity, but the differentiation of *Homo sapiens* as a distinct species, did not come about until perhaps the late Pleistocene. When man actually appeared on the scene, even the youngest mountains were thus already 'as old as the Earth' for him.

This explains his total lack of distinctive high altitude specializations; he is not an integral product of the high altitude ecosystem, but merely an intruder in it. As an intruder, man alters the ecosystem and disturbs the harmony and also finds himself in disharmony. *Even 'perfect' acclimatization, produced by prolonged residence of centuries on high mountains, is basically a pathological state and would by no means entitle him to be considered as a hypsobiont animal.*

5. References

1. Agadzhanyan, N. A. & G. Machinskii, 1968. Vlyanie dlitelnoi gipoksinezii na visotnuyu ustoichvost belykh krys. *Kosm. Biol. Med.*, 2 (1): 25–28.
2. Agadzhanyan, N. A., M. M. Mirrakhimov, V. A. Isabaeva, V. K. Vasilev, G. A. Davydov, Zh. S. Dubinina, A. F. Zavodovskii & I. Kovaleva, 1968. Osobennosti izmenenniya gaszoobmena u gortsev i u zhitelei ravininy pri vypolnenii fizicheskii nagruzki. *Sb. nauch. Tr. Kirg. Med. Inst.*, 47: 17–22.
3. Alexander, A. F., D. H. Will, R. F. Grover & J. T. Reeves, 1960. Pulmonary hypertension and right ventricular hypertrophy in cattle at high altitude. *Amer. J. vet. Res.*, 21 (81): 199–204.
4. Altland, P. D., B. Highman & B. D. Nelson, 1968. Serum enzyme and tissue changes in rats exercised repeatedly at altitude: Effects of training. *Amer. J. Physiol.*, 214 (1): 28–32.
5. Amor, H., E. Humpeler & P. Deetjen, 1973. Untersuchungen auf Veränderungen der Sauerstoffaffinität des Hämoglobins in mittleren Hohenlagen (2000 m). *Wien. kli. Wochenschr.*, 85 (42/43): 700–702.

6. Barbashova, Z. I. 1960. Specific and nonspecific lines of acclimatization to hypoxia. Acad. Sci. USSR. Problems of Evolution of Physiological Functions. (Transl.)

7. Barbashova, Z. I. 1969. Dynamics of adaptive reactions at cellular level during training of rats to hypoxia: Review of research. *Intern. J. Biometeorol.*, 13 (3/4): 211–217.

8. Barbashova, Z. I., G. I. Grigoreva, V. V. Ermilova & Z. G. Fomia, 1959. Influence of the nervous system on hypoxic erythrocytosis. *Fiziol. Zhur.*, *USSR*, 45 (7): 103–111 (Transl.).

9. Brauer, R. W., D. E. Parrish, R. O. Way, P. C. Prett & R. L. Pressotti, 1970. Protection by altitude acilimatization from exposure to oxygen at 835 mm Hg. *J. appl. Physiol.*, 28 (4): 474–481.

10. Campbell, J. A. 1935. Further evidence that mammals cannot acclimatize to 10% oxygen or 20,000 ft altitude. *Brit. J. exptl. Pathol.*, 16: 39.

11. Dejours, P., R. H. Kellogg & N. Poace, 1963. Regulation of respiration and heart rate response in exercise during altitude acclimatization. *J. appl. Physiol.*, 18 (1): 10–18.

12. Faulkner, J. A., J. T. Daniels & B. Blake, 1967. Effects of training at moderate altitude on physical performance capacity. *J. appl. Physiol.*, 23 (1): 85–89.

13. Gill, M. B. & L. G. C. E. Pugh, 1964. Basal metabolism and respiration in men living at 5800 m (19,000 ft). *J. appl. Physiol.*, 19 (5): 949–954.

14. Hansen, J. E., J. A. Vogel, G. P. Stelter & C. F. Consolazio, 1967. Oxygen uptake in men during exhaustive work at sea level and high altitude. *J. appl. Physiol.*, 23 (4): 511–522.

15. Heath, D. & D. R. Williams, 1977. Man at High Altitude: The pathophysiology of acclimatization and adaptation. Edinburgh: London: New York. Churchill Livingstone.

16. Hultgren, H. N., J. Kelly & H. Miller, 1965. Pulmonary circulation in acclimatized man at high altitude. *J. appl. Physiol.*, 20 (2): 233–238.

17. Hultgren, H. N., J. Kelly & H. Miller, 1965. Effect of oxygen upon pulmonary circulation in acclimatized man at high altitude. *J. appl. Physiol.*, 20 (2): 239–243.

18. Hultgren, H. N., W. S. Pickard & C. Lopez, 1962. Further studies of the high altitude pulmonary oedema. *Brit. Heart J.*, 24 (1): 95–102.

19. Hunt, J. 1956. The Ascent of Everest. London: Hodder & Stoughton, pp. 1–300.

20. Kaulhersz, J. 1968. Niektore atualne zagadnienia fiziologii vysokogorskiej. *Folia med. Cracova*, 10 (3): 377–396.

21. Klausen, K., D. B. Bill & S. M. Horvath, 1970. Exercise at ambient and high oxygen pressure at high altitude and at sea level. *J. appl. Physiol.*, 29 (4): 456–563.

22. Lahiri, S., F. F. Kao, R. Velasquez, C. Martinez & W. Pezzia, 1970. Respiration of man during exercise at high altitude: Highlander vs. lowlander. *Resp. Physiol.*, 8 (3): 361–375.

23. Mani, M. S. 1968. Ecology and Biogeography of High Altitude Insects. The Hague: Dr W. Junk, Publishers.

24. Mani, M. S. 1974. Fundamentals of High Altitude Biology. New Delhi: Oxford & IBH. Publishers, pp. 196.

25. Mani, M. S. 1077. Heredity and Evolution. New Delhi: Oxford & IBH Publishers.

26. Miles, W. R. 1964. Oxygen consumption during the yoga-type breathing patterns. *J. appl. physiol.*, 19 (1): 75–82.

27. Moncola, F., A. Carcelen & L. Betea, 1970. Physical exercise, acid-base balance and adrenal function in newcomers to high altitude. *J. appl. Physiol.*, 28 (2): 151–155.

28. Monge, C. 1948. Acclimatization in the Andes. Baltimore: Johns Hopkins Univer. Press.

29. Monge, C. M. 1955. El concepto de acclimatacion. *An. Fac. Cien. Med. Lima*, 38 (1): 1–8.

30. Murray, R. H., S. Shropshire & L. Thompson, 1967. Attempted acclimatization by vigorous exercise during periodic exposure to simulated altitude. *US Air Force Tech. Doc. Rep.*, Amr. Tr., 67–114, 1–11.

31. Pugh, L. G. C. E. 1967. Athletes at altitude. *J. Physiol.*, 192 (3): 619–646.

32. Rizzo, P. J. 1968. Altitude e trabalho fisico. *Rev. Brasil. Med.*, 25 (8): 567–569.

33. Saltin, B. 1967. Aerobic and anaerobic work capacity at 2300 metres. *Med. Thorac.*, 24 (4): 205–210.

34. Sergienko, A. V. 1970. Kompleksoe vliyanie na organism zhivotnikh ostroi gipoksii i vysokoi temperatury okruhayushchei sredy. *Kosm. Biol. Med.*, 4 (3): 13–18.

35. Severinghaus, J. W., W. A. Mitchell, R. W. Richardson & M. M. Singer, 1963. CSF pH in

man during acclimatization to high altitude. *47th Ann. Meetg. Fed. Amer. Soc. exptl. Biol. Fed. Proc.*, 22 (2): 223.

36. Shataline, A. S. & A. N. Potachuk, 1968. Premenenie fizicheskii uprazhneii v usloviyak vysokogorya Pamira kak faktora aktivnoi akklimatizatsii. *Nauch. Tr. Tashkent. Univ.* 338: 124–127.

37. Slonin, A. D. 1957. Znacheni prirodnykh faktorov unesheni sredy dlya fiziologii selskokhzyaistvennykh zhivotnykh. Akad. Nauk SSSR, 173–179. *Zhur. Biol.*, 1959.

38. Syrotonin, N. N. 1965. Pro rizni varanty akklimatyzatsiyi do vysokhinoko klimatu. *Fiziol. Zhur. Akad. Nauk. UKSSR*, 11 (3): 283–288.

39. Weihe, W. H. 1967. Effects of ambient air temperature on the acclimatization of rats to high altitude. *Proc. 3rd intern. Biometeorol. Congr.*, 1963, *Biometeorol.*, 2 (1): 219–225.

40. West, J. B. 1963. Work limitation at very high altitude. *J. appl. Physiol.*, 167 (1): 14.

41. West, J. B. 1967. Exercise limitations at increased altitude. *Med. Thorac.*, 24 (6): 337–337.

7 For the future

L. E. Giddings & M. S. Mani

The highlands have always presented a most fertile and interesting field for research. Still, with the exception perhaps of some recent work in human biology, most studies have been motivated by the personal, academic interest of individual researchers. Indeed, the motivations of the three authors of this book have been quite diverse, ranging from a purely academic interest in highlands entomology, to the more practical problems of medicine and laboratory operations in the inhabited highlands.

Regardless of motive, there is a wide range of studies that need to be made, much more than in lowlands areas. In addition to their academic interest, more and more of them are becoming important to our comprehension of our world as a whole. They are also vital for the scientists and engineers in highland countries and national programmes of development.

This chapter presents some ideas on the state of highland studies and, when possible, suggests specific directions for further work. No special distinction is made between the theoretical and the strictly practical tasks, since the line between them is rather fine. The order of subjects roughly follows the arrangement of this book, chapter by chapter, with miscellaneous topics added at appropriate places. Many additional problems of studies would naturally suggest themselves to the reader.

1. The Physical Environment

Probably the greatest need for attention is the measurement and understanding of the physical environment of some of the major highland regions. The parameters listed in Chapter II are only approximations, mostly derived from measurements of free upper air. The surface conditions on highland ground are certainly different.

At a given latitude, the surface air temperature is not the same as the free upper air temperature at the same altitude, even on the average. Surface heating and cooling phenomena cause large variations from the average values given in the standard atmospheres. There is no reason to expect that heating will, even on the average, offset cooling to produce the free air temperature at high altitude.

It is not as apparent that the surface barometric pressure is likely to deviate from the standard atmospheres, but significant variations have been observed (12). In any case, even for the free upper air, the standard atmospheres are strictly valid only for the northern hemisphere. Studies of the actual surface barometric pressure, as· a function of altitude on land surfaces, are urgently needed at least on different major highland regions.

An attempt to construct standard surface atmospheres for several latitudes would be very instructive; they should include, as a minimum, both temperature and pressure. Note that Fig. 15-1, derived from the standard atmospheres, shows that pressure varies strongly with latitude at high altitude (12). Still, it is not apparent how highland surface pressures will vary.

Most data for the standard atmospheres, and indeed, most published observations on the physical environment, refer to the northern hemisphere. For lack of better data, we have been forced to assume that they are equally valid for the southern hemisphere. This is certainly not completely true, given that the distribution of land and water in the hemispheres are different (Chapter I). The highland environments of the two hemispheres need to be compared in greater detail than at present and this will require numerous observations.

The above comments refer to the overall physical environment, or average conditions, in the sense of the standard atmospheres. The fine details of highlands also need to be measured and described. The surface variation of physical parameters for any given country or region is a very valid subject for study.

Finally, some attention needs to be paid to the problems of meteorological stations in the inhabited highlands, and the measurement of meteorological conditions by mountaineering expeditions. Perhaps review articles could be prepared, integrating experiences at cold stations, at isolated stations, and on expeditions, as guidelines for those with less experience of the problems of making meteorological measurements under highland conditions.

It must be painfully evident from Chapter VI that precision barometry for the highlands has never been addressed in print, at least to the knowlege of the authors. Although in principle, one needs only to follow the existing guidelines of the WMO, there is little doubt that establishing a reference barometer in a highlands laboratory is bound to present several practical problems. A highland scientist or science administrator must be able to find out how to set up a precision barometer and be assured that it is functioning properly, with the confidence that the instructions are valid for his special conditions. This can only be said to be true on the basis of actual future investigations in highlands.

The problem of boiling point correction for the lower pressures of the highlands is quite well treated in literature at present. Still, the formulations that appear best were made for the requirements of boiling under artificially lowered pressures. The choice of formulation for practical work in a highlands laboratory, and especially in the teaching laboratory is not available at present. Perhaps the German chart is the easiest to use for known compounds, but we need specific formulation that would be most convenient for both known and unknown substances.

For measuring humidity, the wet- and dry-bulb thermometric technique is not the easiest method, and it may well not be the best. One needs to know how the use of a sling psychrometer compares with dial devices, such as the hair and Durotherm element instruments. Theory suggests that they may be reasonably precise and, at the same time, independent of altitude. Well

226

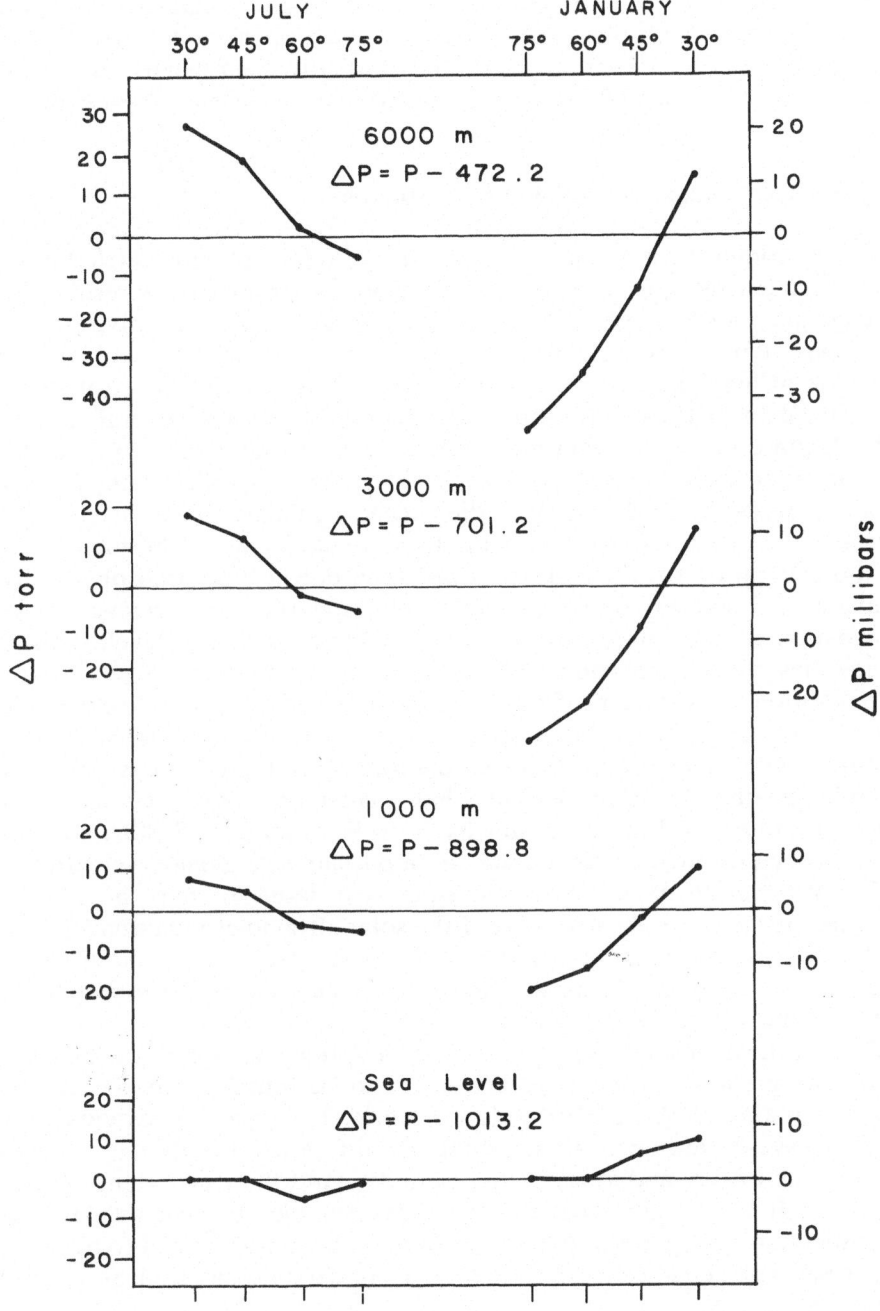

Fig. 15-1. Variation of pressure with latitude.

planned mobile experiments are needed to compare ventilated psychrometers of several types, dial psychrometers, recording hygrographs, dew-point devices and perhaps others. Dew point or chemical devices might well be used as standards. Experiments should naturally be repeated at several different altitudes, varied temperature conditions, different seasons, etc.

2. The Electromagnetic Radiation Environment

The most comprehensive and recent studies of solar radiation carried out so far have been integral parts of various space programs. As a result, their findings are mostly devoted to the upper air and space, the environments most important for rockets and satellites. In no case do they deal with the radiation at highland surfaces. By the same token, the fine classical studies were made in just a few locations. We cannot know how general they are until they are made for other highlands areas. To what extent, for example, are studies from the rugged Alps applicable to the high, flat plains of other highland areas in South America, the rugged Himalaya, Tibet, etc?

In Chapter III reference was made to a modification of Elterman's work on the estimation of attenuation of radiation due to the atmosphere. This estimate is, however, based on the assumption that the average aerosol distribution at highlands surfaces was the same as at sea level. This, of course, has never been confirmed and needs to be studied.

In addition to average profiles of atmospheric aerosols, specific cases need to be better understood than at present known. How do non-industrial areas compare with industrialized areas; deserts, with humid areas; areas of ascending, with those of descending winds? These are not of great interest to space programs, and therefore not likely to be studied by NASA or other space agencies, but they are extremely important to highlands scientists.

The distribution of ozone is now a pertinent issue for the whole world, since the ozone layer absorbs most of the solar ultraviolet radiation (31, 38). It is now the subject of worldwide studies from satellites. As these studies reach fruition, we should have a better understanding of its hemispherical and seasonal variations.

The practical problem of solar energy resources is rapidly coming to a head. The great solar resources are surely to be found in the deserts and highland plains of the underdeveloped world. But just how extensive are these reserves? One needs to know the quantity, the intensity, the distribution in time of day and season, etc., to understand the true extent of these resources. It is becoming apparent that ways will soon become available for the direct harnessing of this immense resource. Inventories need to be made, and these will need to use information from satellites as well as ground-based stations.

As with the other environmental parameters, there is a great need for regional and local measurements. The methods and discussions presented in this book are only useful as guides. The greatest need is for actual measurements. For example, the diffuse ultraviolet radiation is obviously high in the highlands, as every sun-tanned traveller who has kept out of the sun can testify. But just how high is it, and how does it vary from place to place? We have at present no answers to these questions. Actual measurements are urgently needed. There is little doubt that photography changes with increasing altitude. At sea level the colour temperature of direct sunlight is about 5000°K, and that of indirect radiation is more nearly 12,000 to 27,000°K, due to strong scattering of blue light. It is obvious that all these figures will change with altitude. In fact, changes are apparent even in amateur snapshot photography. All factors involving photography need to be quantified, including the necessity of special filters, reduction of exposure, etc.

3. Corpuscular Radiation

The distribution of corpuscular radiation outside of the Earth's atmosphere became known only as a result of the space age. Only with satellites and rockets could the solar wind, the flow of particles from the sun, be mapped. Radiation from other sources were also known only as a result of experiments made in the outer space.

The distribution of corpuscular radiation over the Earth is known to be complex, but little attention has been paid to the highlands. We need to know, for example, if there is an appreciable increase in corpuscular radiation in the highlands. If so, is it of a sort that has biological significance? Is it a significant environmental parameter, more so than in lowlands?

4. Chemical Effects

The chemical effects of the highlands, the indirect results of the lower pressure, can be understood quite easily. As shown in Chapter V, the practical problems can be refractory. In particular, compensating chemical reactions of organic substances can be tricky. The mathematical model presented there, which serves so well for inorganic reactions, also shows that compensating for high altitude by longer boiling is often not only impractical, but does not also yield the desired result.

A consequence of this chemical effect that has far-reaching consequences is the sterilization, say of canned goods or medical instruments, by the heat of boiling water. The destruction of living organisms is the basis of sterilization, but for such complex substances, the increase in time of boiling suggested by the mathematical model is not always a practical way to compensate for the reduced pressure of the highlands.

The practical problems of sterilization under highland conditions need to be studied in detail. Although a proper autoclave is largely unaffected (provided that gauges actually measure temperature instead of inferring temperature from pressure), open boiling of instruments is problematical.

This most widely used method of sterilization needs to be studied as a function of altitude. One needs to know how long instruments need to be boiled at any given altitude to ensure sterilization. It would be comforting to have such studies made in actual highlands areas, rather than in reduced pressure chambers in laboratories at sea level.

As a practical matter, many autoclave and pressure cookers for the most diverse application actually measure pressure but report temperature. The actual gauges are usually calibrated 'gauge' instead of 'absolute', and actually measure *differences* from the local atmospheric pressure, although their dials give the absolute readings that would be correct at sea level. As a result, neither apparent pressures nor temperatures of such gauges are correct in highland areas, and their error increases with altitude. Following the common instructions for sterilizing canned goods has resulted in having on hand unsterilized food in an accident known to one of us (LG). The potential problems are great. For purely practical reasons, this subject needs practical attention.

5. For the Engineer

Practical attention needs to be given to highlands problems, likely to be faced especially by chemical and mechanical engineers. the theoretical understanding of azeotropes that currently exists, for example, must be translated into practical rules for the chemical engineer. The effects of altitude on internal combustion engines is very complex, and needs to be studied in a practical way for the mechanical engineer. What can one do to an automobile, for example, to ensure that it will function well at high altitudes? All engineers need to know how materials deteriorate under the increased ultraviolet radiation environment of the highlands.

Reference has already been made to the inventory of solar resources, using, in part, satellite technology. Since the highland-plains contain large, unused areas, they also have potential for the use of indirect solar energy in the form of wind power. Again, measurements of the available energy need to be made by experiment. This may also be feasible in mountainous areas; this also needs to be investigated.

6. Biology

Highland biology is a relatively new field, with almost unlimited scope for future work. Vast areas of highlands in all parts of the world have still to be explored systematically by botanists and zoologists. With perhaps the notable exception of insects from the Alps and parts of the Himalaya, all other groups of organisms have almost been completely neglected by biologists so far. Even in this group urgent taxonomic revisions are badly needed. There are also a number of intriguing problems in the high altitude insect life which need clarification by future investigations. Programmes of studies on the vegetation, flora, fauna, general ecology and biogeography of some of the major highland regions should prove extremely productive. Useful vegetation and ecology maps of the Alps and small parts of the Nepal Himalaya have recently been published. Such maps of the rest of the Himalaya and other principal highland regions should prove immensely useful.

The use of space technology and appliances in studies of highlands problems is just in its infancy. Some space-related items that are particularly promising are vegetation inventories from Landsat images; vegetation and ecological zone discrimination of remote areas from meteorological satellites; inventory of solar resources; measurement of parameters useful for agriculture, such as the hydrological cycle, cloud cover, and perhaps insolation; extension of surface air temperatures from sparse networks; the preparation and use of ecological, and perhaps agricultural data bases in the efficient image formats; and the measurement of solar energy resources.

7. Man in Highlands

Two major aspects of the problems of man in highland have to be investigated by future workers, viz. (i) human comfort and health and (ii) the consequences of human impact on highland ecosystems.

Though the problem of human acclimatization to high altitude regions has been studied by a number of workers in the past, even now we really have no satisfactory answers to many basic questions. A whole series of questions relating to the problem of high altitude acclimatization has still to be carefully studied in actual highland areas, rather than in a laboratory situated in lowland regions. The rôle of physical activity, clothing, food, etc. in human acclimatization is at present only very imperfectly understood. There is, therefore, urgent need for careful clinical and pathological studies in some of the major inhabited highland areas. No doubt the valuable knowledge which we have accumulated from studies in aviation medicine would prove useful, but the data have to be carefully tested and suitably modified for actual application in highlands.

Many studies have determined the variables of human adaptation and comfort in extremes of climate but mainly under arid conditions near sea level (43). The heat balance, performance at work and other factors, have been measured under extremes of hot and cold, aridity and humidity, and other factors, but few have considered the profound modifying effects of altitude. In particular, the appropriate temperature and humidity for comfort could be established. Many interesting data can probably be gleaned from studies of men in the aerospace environment (11, 32, 37), but safety in highland laboratory needs to be studied in a systematic way as a function of altitude. The safety engineer and particularly technicians charged with safety matters, have very few quantitative materials which they can now use in the highlands. Conversion of diverse available data to highlands conditions is difficult even for the professional and certainly impossible for the working technician. A reasonable first start could consist of the measurement of flash points in the highlands.

Closely linked with human comfort is the problem of changes in cooking methods to be followed under highland conditions. Some valuable studies have been made in this direction at lower elevations (8, 9, 15, 26). Many aspects of cooking at 1500 m, have been studied at the Colorado State University, and a few have been studied at 3000 m. At the Mountain Research Station of the University of Colorado water boils at 88°C (190°F). "Potatoes and vegetables such as brussel sprouts take about an hour to be cooked unless cut up into very small pieces. Prepared foods such as 'instant' rice require about twice the suggested quantity of water. In contrast, roasts, fried foods, and for some strange reason, pre-made pancake mixes need no adjustments in cooking methods." The data obtained so far, while no doubt valuable, need to be very considerably modified by actual experimental investigations, before they can serve as useful guides for inhabitants of higher elevations than Colorado, for example the Tibet Plateau and the Bolivian Altiplano, living at elevations higher than 4000 m. Cooking needs to be studied at the very high altitudes found in the habitable highlands, and the very high cities of these areas would make attractive laboratories. Experiments are needed on baking, on candy making, cooking by boiling of several kinds of foods and others.

It is generally recognized that highland ecosystems, which had in the past remained undisturbed by man, are currently being subjected to the impact of intense human pressure. The general ecological balance in most highland regions is being increasingly disturbed and often irrevocably upset (2, 4, 5, 6, 7, 10, 18, 21, 27, 29, 30 & 40). The major sources of these disturbances in highland ecosystems may eventually be traced back to drastic changes in the traditional land use practices (14). The complex factors which have led to these changes include population growth (2, 19, 23), emigration of native inhabitants and heavy influx of outside peoples, unlimited demands of tourism (35), extensive deforestation and logging, abandonment of traditional mountain farming and modernization of agricultural practices, over-grazing and extension of pasture into forest covered land, military operations and technological developments like quarrying, mining, tunnelling, hydro-electric installations, etc. It is known, for example, that under the impact of heavy demands of tourism mountain farming has diminished by 15% in the Austrian Alps during the past twenty years, by 25% in the Bavarian Alps and by 20% in the Italian Alps within ten years (2, 13). In all these areas, prime agricultural land is increasingly being developed as building sites for hotels. These changed conditions have affected the resources potential first and at the same time increased our losses. While tourism lies at the root of ecosystem disturbances in the Alps (4), it is recent invasion of modern technology that is the principal factor in population explosion and massive ecological upsets in the Himalaya (29). In certain remote inhabited Himalayan high valleys, prior to the introduction of technological developments, the natives followed age-old methods of cultivation in stone-reinforced terraces and the population was only 13,000 (Plate 12-3).

232

Further, the population growth during fifty years was no more than 2000. Rapid socio-economic changes, undreamt of prosperity and population explosion came all-too-suddently in the wake of invasion by modern technology. Events outside India, particularly the occupation of Tibet by China and the subsequent attacks by China and Pakistan on India revealed the strategic importance of these slumbering highland areas. Soon followed the establishment of air-fields, construction of modern roads, etc. as parts of defence developments. These developments at once opened up the remote and formerly almost inaccessible areas to the outside world and then inevitably followed rapid modernization of transport, mechanization of agriculture, hydro-electric installations, programmes of health, education, heavy tourist traffic, etc. Within two decades between 1951 and 1971 the population more than doubled. As a direct result of these events, the forest covered area has shrunk by more than 70% of its former size and building sites are continually expanding.

We know that most highland regions are theatres of periodical disasters and hazards like soil erosion (27, 39), landslides and slips, avalanches (3), earthquakes, torrents, floods, etc. In the past years, when the human interference with natural ecosystems was neither so widespread nor intense, all disasters were due entirely to natural causes and very little indeed to human negligence and mismanagement. In recent years, however, under the impact of man, the frequency and intensity of such disasters have greatly increased. Man is also increasingly venturing into disaster prone areas. It is reported, for example, that at least 66% of the disasters affecting man in the Austrian Alps in recent years is directly to be traced to human activities (4). As compared to 1945–1950, there has been a four-fold increase in disasters during 1963–1968. In some parts of the Tyrolean Alps the frequency of floods and avalanches has increased by 240–968%. This is directly the result of destruction of forest by 63.8% to meet the demands of tourism. Of the diverse harmful consequences, loss from soil erosion is perhaps one of the most serious. Normally soil erosion is a slow process in balance with soil formation, but when vegetation cover is destroyed the rate of erosion is greatly accelerated (5, 6, 18, 21, 22, 27, 39). Deforestation is also correlated with a series of other harmful changes (7, 10, 17, 20, 24, 25, 28 & 30). The seriousness of the consequences of human impact on highland ecosystem may readily be appreciated by the increasing attention paid to disaster prevention measures in some European countries (4). The measures so far undertaken in these countries include registering the threatened areas, forecasting the natural disasters, trying temporary disaster prevention measures as well as permanent protection by afforestation (1, 4). The results obtained so far serve merely to emphasize the need for further intense studies. In nearly all highland regions there is most urgent need for careful 'pre-impact surveys' of selected areas. Such surveys should help in determining the basis for environmental disturbances and for identification of vulnerable areas that could serve as indicators of environmental stress. The surveys should cover mapping of vegetation, distribution of wild life, peculiarities of geology, soil and relief, meteorology and also archaeological sites. Besides these surveys carried out from the ground, much valuable data can be obtained by the application of remote-sensing technique. Studies of the environmental response to human impacts should also be made. Such studies should identify the threshold values of changes in the ecosystem, so that the critical point beyond which recovery is difficult or is not possible, may be forecast.

Urgent research is also required on the means of reducing the technological impact. For example we need comparative studies on the ecology of dam sites and hydro-electric projects both before and after the impounding of water, studies on the transportation and processing of mining material suggesting reduction of disturbance to the local inhabitants and to visitors, studies on the cumulative effects of fertilizers and pesticides in highland soil and the long-term effects of mechanization of farms, etc. We also need studies on the rehabilitation of highland areas, which have been badly disturbed by human activity. Similarly one would like also studies on the rehabilitation of areas which are suffering ecologically from the effects of emigration of highland natives. Another important line of work lies in areas in which natural calamities are known to have increased in recent years as a result of human activity. Work is also needed for effective management practices of national parks.

Ecologists and human biologists and social scientists should cooperate in order to incorporate human populations into the ecosystem approach so as to develop a common frame for integrated studies of physical, biotic and socio-economic aspects of the whole highland ecosystem. Within the ecosystem and human community, these studies should consider the essential resource requirements and flow, degree of imbalance in the ecosystem, potential for increase or fall in resources exploitation and requirements for maintenance of viable social systems. Socio-cultural studies should deal with human values to static and changing life styles. Human biologists may relate these to health and population dynamics. Ecologists will have now basis for comparison of the relative degree of perturbation in ecosystem and possibility of achieving homeostasis in ecosystem.

The ideal condition to be aimed at is a beneficial combination of development of the natural resources and human settlements that would improve the quality of human life in the highland habitats, without at the same time damaging the life-supporting capacity of the highland ecosystems.

8. References

1. Aulitzky, H. 1955. Die Bedeutung meterologischer und kleinklimatischer Untersuchungen für die Aufforstung im Hochgebirge. *Wetter. Leben*, 7: 241–252.
2. Aulitzky, H. 1971. Sensitive mountain regions. Council of Europe, European Committee for the Conservation of Nature and natural resources, ad hoc study group on soil conservation. CE/Nat.71.54. Strasbourg.
3. Aulitzky, H. 1972. Study of the causes of avalanches with the abandonment of agricultural and pastoral activities. Supplement 3.
4. Aulitzky, H. 1974. Endangered alpine regions and disaster prevention measures. *Eur. Commt. Conserv. Nat. nat. Res.*, Council Europe, Strasbourg pp. vi + 103 + 12.

5. Bailey, R. W. 1935. Epicycles of erosion in the valleys of the Colorado Plateau Province. *J. Geol.*, 43: 337–353.
6. Bailey, R. W. 1941. Land erosion – Normal and accelerated in semi-arid West. *Trans. Amer. geophys. Union.*
7. Bormann, F. H. 1971. In: Mathews *et al. Mans Impact on Terrestrial and Oceanic Ecosystems. Population, Technology and diminished man.*
8. Bowman, Ferne, M. Frease, W. Dilsaver & P. Splittgerber, 1971. 'Making Yeast Breads at High Altitudes', Bull. 526S; Bowman, Ferne, W. Dilsaver & M. Frease, 'Wheat-Gluten-Egg and Milk Free Recipes for Use at High Altitudes and Sea level', Bull. 544s (91970).
9. Boyd, Margaret S. & Mayme C. Schoonover, 1965. 'Baking at High Altitude', Bull 427, Agricultural Experiment Station, University of Wyoming, Laramie, Wyoming 82071, U.S.A.
10. Coleman, E. A. 1953. Vegetation and Watershed Management: An apprisal of vegetation management in relation to water supply, flood control and soil erosion. New York: Ronald Press Co. pp. 1–412.
11. 'Compendium of Human Responses to the Aerospace Environment', esp. vol. 1, sect. 6 and vol. 3, sect. 10 and 12, Report by the Lovelace Foundation for Medical Education and Research, NASA Contractor Report CR-1205, 4 Volumes, Washington D.C. (1968).
12. ESSA, NASA, USAF, 'US Standard Atmosphere Supplements, 1966', U.S. Government Printing Office, Washington D.C.
13. Franz, H. & C. S. Holling, 1974. Alpine areas workshop. *Internat. Inst. appl. Syst.* Analysis Conf. Proc. Laxenburg (Austria) May 5, 1974 pp. 1–80.
14. Graham, E. H. 1944. Natural Principles of Landuse. New York: Oxford Univ. Press, pp. 274.
15. Greenland, David 1977. Living on the 700 Millibar Surface. The Mountain Research Station of the Institute of Arctic and Alpine Research, *Weatherwise*, 30, 233 (*December*).
16. Guntschl, E. 1970. Hochwasserschutz and Raumordnung, Schriftreihe Österr. Ges. Raumforsch. Raumplannung, Wien, 11.
17. Hepting, G. H. 1971. Climate and Forest Diseases. 1962. In: Mathews *et al.* Man's Impact on Terrestrial and Oceanic Ecosystems. MIT Press, pp. 203–226 (See also *Proc. V. World Forrestry Congr.*, 2: 842–847.
18. Horton, R. E. 1945. Erosional development of streams and their drainage basins. Hydrophysical approach to quantitative morphology. *Bull. geol. Soc. America*, 56: 275–370.
19. Hulett, H. R. 1970. Optimum World Populations. *Bioscience*, 20 (3): 160–161.
20. Kaplan, L. D. 1960. The influence of carbon dioxide variations on the atmospheric heat balance. *Tellus*, 12: 204–208.
21. Karl, J., W. Danz & J. Mangelsdorf, 1969. Der Einfluss des Menschen auf die Erosion in Bergland. Schriftreihe der Mayrischen Landsstelle für Gewässerkunde. Münnchen.
22. Knapp, R. T. 1941. A concept of the Mechanics of the erosion cycle. *Trans. Amer. Geophys. Union*, 1941: 255–257.
23. Ketchum, B. H. 1971. Population, natural resources and biological effects of pollution of estuaries and coastal waters. In: Mathews *et al.* Impact of Man on Terrestrial and Oceanic Ecosystems. NIT pp. 59–79.
24. Landsberg, H. E. 1958. Trends in Climatology. *Science*, 128: 749–758.
25. Lieth, H. 1963. The rôle of vegetation in the carbon dioxide content of the atmosphere. *J. geophys. Res.*, 68: 3887–3898.
26. Lorenz, Klause & W. Dilsaver, 1972. 'Mile-High Cakes for the 70's', Bull. 556S and several others, all from the Colorado State University, Fort Collins, Colorado 80521, U.S.A.
27. Lowdermilk, W. C. 1935. Acceleration of erosion above geologic norms. *Trans. Amer. geophys. Union*, 15th Ann. Meetg., pp. 508–509.
28. MacDonald, G. F. J. 1969. The modification of the Planet Earth by Man. *Technology Rev.* (Oct.–Nov.).
29. Mani, M. S. & S. Singh, 1978. Impact of Man on Mountain Ecosystems in India. Rep. Seminar. *Mem. School Ent.*, 6: 1–128.

235

30. Mathews, W. H., F. E. Smith & E. D. Goldberg, 1971. Man's Impact on Terrestrial and Oceanic Ecosystems. Cambridge: Mass. MIT Press. pp. xiv–540.
31. Mitchell, J. M. Jr. 1961. Recent secular changes of global temperature. *Ann. N.Y. Acad. Sci.*, 95: 235–250.
32. Parker, James F. Jr. & Vita R. West, 1973. 'Bioastronautics Data Book', 86–106, NASA SP-3006, Washington.
33. Peterson, E. K. Carbon dioxide affects global ecology. *Environ. Sci. & Tech.*, 3: 1162–1169.
34. Plass, G. N. 1959. Carbon dioxide and climate. *Sci. Amer.*, 201: 41–47.
35. Poncet, M. A. 1969. Tourisme et réstuaration et conservation des terrain en montagne. Relation entre l'amenagement des bassins versants de Montagne et le tourisme. pp. 244–260.
36. Poncet, M. A. 1971. Situation de la torrentialité dans les Alpes francaises – Problemes noureaux diés à l'évolution de l'économie régionale et du paysage végétal. Orientations nouvelles dans la lutte contre l'érosion torrentialle. Internationales Symposium. 'Interpraevent 1971'. Villach. Grenzen und Möglichkeiten der Vorbeugung der Unwetterkatastrophien. 2: 311–331. Klagenfurt.
37. Roth, Emmanuel M. 1967. 'Space-Cabin Atmospheres', part IV, Engineering tradeoffs of one- versus two-gas systems', NASA SP-118, Washington.
38. Russel, R. J. 1941. Climatic changes through the ages. Yearbook Agr. U.S. Dept. Agr. Washington DC 1941: 7–9, 67–97.
39. Sharpe, C. F. S. 1941. Geomorphic aspects of normal and accelerated erosion. *Trans. Amer. geophys. Union*, pp. 236–240.
40. Thomas, Jr. W. L., C. O. Sauer, M. Bates & L. Munford, 1970. Man's role in changing the Face of the Earth. Chicago: Chicago Univ. Press pp. xx + 1193.
41. Watt, K. E. F. 1968. Ecology and Resource Management. New York: McGraw-Hill Book Co.
42. Wenger, K. F., C. E. Ostron, P. R. Larson & T. D. Rudolf, 1971. Potential effects of global atmospheric conditions on forest ecosystems. In: Mathews *et al.* Man's Impact on Terrestrial and Oceanic Ecosystems. MIT Press, pp. 192–202.
43. Wyndham, C. H. 1964. Performance and Comfort Standards in Relation to Climate, 189–198 of Environmental Physiology and Psychology in Arid Regions, 'Proceedings of the 1962 Lucknow Symposium, UNESCO.

Index

237

242

248